SCIENCE IN WONDERLAND

SCIENCE in WONDERLAND

The scientific fairy tales of Victorian Britain

MELANIE KEENE

OXFORD

UNIVERSITY PRESS

OXFORD
UNIVERSITY PRESS

Great Clarendon Street, Oxford, OX2 6DP,
United Kingdom

Oxford University Press is a department of the University of Oxford.
It furthers the University's objective of excellence in research, scholarship,
and education by publishing worldwide. Oxford is a registered trade mark of
Oxford University Press in the UK and in certain other countries

© Melanie Keene 2015

The moral rights of the author have been asserted

First Edition published in 2015

Impression: 1

Published in the United States of America by Oxford University Press
198 Madison Avenue, New York, NY 10016, United States of America

British Library Cataloguing in Publication Data
Data available

Library of Congress Control Number: 2014950221

ISBN 978–0–19–966265–4

Printed in Great Britain by
Clays Ltd, St Ives plc

Here about the beach I wander'd, nourishing a youth sublime
With the fairy-tales of science, and the long result of Time.

Tennyson, 'Locksley Hall' (1835)

ACKNOWLEDGEMENTS

My thanks to all at Oxford University Press for their help in seeing this work into print, particularly Latha Menon, Emma Ma, and Jenny Nugee. Thanks to Sophie Basilevich for her help with the images.

Also many thanks for reading and commenting on parts of this manuscript to Jim Secord; members of the History of Science Workshop in the Department of History and Philosophy of Science, University of Cambridge; Amy Blakeway, Daniel Trocmé-Latter, and Alice Wilson at Homerton College, Cambridge; and Marie Besnier. Feedback from audiences at conferences of the British Society for the History of Science, the British Society for Literature and Science, the Cambridge Children's Literature Seminar, the Oxford History of Childhood Seminar, the Commission for Science and Literature, and participants in an Aberdeen workshop on literature and science was also invaluable.

As ever, I could not have written this book without the love and support of Nick and my family.

CONTENTS

LIST OF ILLUSTRATIONS

LIST OF PLATES

1. The 'Crystal Palace' from the Great Exhibition installed at Sydenham. Wellcome Library, London.

2. 'Sidney's introduction to the fairy', frontispiece, *Fairy Know-a-Bit*. The Bodleian Library, University of Oxford, M92.G03314.

3. 'Monster Soup commonly called Thames Water', William Heath, 1828. Wellcome Library, London.

4. Cover, Arabella Buckley, *The Fairy-land of Science* (London: Stanford, 1879). The Bodleian Library, University of Oxford, (OC) 198.f.85.

5. 'A Flight to the Moon', from Agnes Giberne, *Among the Stars: Or, Wonderful Things in the Sky* (New York: Robert Carter, 1885), between pp. 172 and 173. The Bodleian Library, University of Oxford, Radcliffe Science Library 1842 e.1.

6. 'A Ride With the Sun', from Lizzie W. Champney, *In the Sky-Garden* (Boston: Lockwood, Brooks, 1877), before p. 105. Courtesy of the New York Public Library.

Nothing but Facts?

We have believed Mr Gradgrind for far too long. When he appeared in 1854, at the beginning of Charles Dickens's novel *Hard Times*, Thomas Gradgrind embodied everything that was dry and dull in Victorian scientific education.[1] Facts alone, he declared, were all that was wanted in life. Children should be able to recall at will the most recondite information about horses' hooves. They should be discouraged from surrounding themselves with anything that was not mathematically exact and accurately representative, even to the very wallpaper of their classroom. Like the jars in Ali Baba's tale, students were vessels to be filled to the brim with useful knowledge. But lying alongside an emphasis on 'facts alone' was a whole range of more fantastical allusions, as Dickens's choice of simile showed: cramming children with information would not 'kill outright the robber Fancy lurking within'.[2] This comparison to a fantastic tale undermined Gradgrind's hyperbolic claim: revealing the author's satirical purpose it referenced his favoured *Arabian Nights*, subtly introducing the message of his book.[3] In this way, Dickens brought out the close relationship between facts and fancy that existed throughout the nineteenth century. Scientific education was not just to be had through fusty educational primers: it was also found in wonderland.

Dickens's novelistic intervention was just one part of a wider educational debate ongoing in the mid-nineteenth century around what children should read and do, and in what ways. The 'rising generation', it was argued, lived in an ever-progressive age, aware of its own innovations and place in history. Novel inventions harnessed the power of steam and electricity; new career opportunities in urban centres and beyond affected societal relationships, working patterns, and more, changing the texture of daily life within Britain, and her Empire across the globe. It was clear—as both Parliamentary committees and the pages of *Punch*, an influential satirical periodical, attested—that the traditional systems of children's education must also 'advance with the age'. Everything from elementary instruction to the contents of the toy-chest, *Punch*'s mock-journalist argued, should be reformed, in order that young audiences best be prepared for the modern world. 'When everything else is moving', it was especially important that provision for children 'should not remain at a stand-still'.[4] The place of the written word was of paramount importance in these debates: print was cheaper than ever before, and children's books and objects—from alphabet guides to dedicated periodicals to table games and toy sets—were a cornerstone of the commercial marketplace (see Fig. 1). The sciences were to be found at the heart of all of these developments, not only as an emblem of the age, but also as a crucial missing element from the curriculum.

Dickens's opening scene in *Hard Times* therefore presented a case-study in taking too literally an extreme reliance on what had been termed 'useful knowledge', intellectual fodder deemed appropriate for both the working classes and the young.[5] Mr Gradgrind, a subscriber to this educational

THE OLD.   THE NEW.

Fig. 1. 'Old and New Toys', *Punch* (1848). One of several illustrations accompanying a satirical article that compared the traditional toys of the nursery with suggested improvements for Victorian Britain. Alongside other depictions of a megatherium rocking-horse and a hot air balloon kite, it recommended replacing a trumpet and toy sword with a blowpipe and an electrical machine.

philosophy, had ensured his children had indeed grown up with 'nothing but Facts'.[6] What followed was a damning indictment of adhering too closely to a way of learning that was scientifically up-to-date, suitably mechanized for an industrial family. The children had, Dickens's narrator disapprovingly noted, 'never known wonder', and therefore had developed no notion of any more fanciful or fairy tale explanation for the features of the natural world. Twinkling little stars were not the basis of nursery songs, but rather objects for scientific analysis: the precocious little Gradgrinds had, the narrator boasted, 'at five years old dissected the Great Bear like a Professor Owen, and driven Charles's Wain like a locomotive engine-driver'.[7] Cows were not to be found in stories and riddles, but were known only as a type of 'graminivorous ruminating quadruped with several stomachs'. Their experience was not just verbal: it was also a practical one:

> The little Gradgrinds had cabinets in various departments of science too. They had a little conchological cabinet, and a little metallurgical cabinet, and a little mineralogical cabinet; and the specimens were all arranged and labelled, and the bits of stone and ore looked as though they might have been broken from the parent substances by those tremendously hard instruments their own names...[8]

This presentation skewered a particular mode of teaching; however, whereas the five Gradgrind children had only grown up with their 'ological cabinets and fact-filled speeches, for many other children in Victorian Britain such hands-on scientific investigations were practised alongside and as part of other types of activities.

Far from being dusty fact-crammers, Victorian science books and articles, shows and activities, artefacts and pastimes, were explicit attempts to combine education with entertainment; or, in the favoured phrase of the day, 'instruction and amusement'.[9] This was not hard to do: the sciences were the most exciting subjects of the century, about which everyone wanted to learn. From miniature marvels to massive monsters, they included a novelty to suit all tastes. The market for scientific amusements boomed throughout the century, as successive fashionable waves broke over the Empire, from fern-collecting to shell-collecting, rock-pooling and the keeping of domestic aquaria, to chemical experiments or night-time star-gazing. Authors and illustrators, manufacturers and publishers, rushed to cater to this upper-middle-class demand for so-called 'rational recreation', claimed as a superior way of spending the leisure hours, itself a nascent concept. However, the desirable balance of entertainment and instruction was tricky to manage, and writers sought to stabilize their works in a range of ways; factual information was couched as conversations, lectures, letters, calendars, journeys, and more. Throughout the century different literary forms were played with and argued for as the best means of writing about and practising the sciences, including for children.[10] Analysing these books, objects, and activities elucidates their content and intent and impact; they were evidently much more than Gradgrindian stereotypes. We will discover that fairy tales might have been made to look a lot like science; but also that science—as Dickens's oriental simile in my opening example reminded us—could look a lot like fairy tales as well.

Science for the people

Many Victorian audiences would have been familiar with Dickens's characterization of attempts to instruct wider publics in scientific subjects. The second quarter of the nineteenth century had been rife with movements to—as it was seen—compensate for both expert and lay deficiencies in scientific understanding. The so-called 'gentlemen of science' had combated a perceived 'decline' in British capabilities with calls for the reform of university syllabi and the defining of new roles for scientific practitioners, and founded the British Association for the Advancement of Science in 1831, whose peripatetic annual meetings would continue throughout the century as a touchstone for current debates, and as fodder for the periodical press.[11] Leading Whig politicians such as Lord Brougham saw the spread of scientific understanding as a cornerstone of their Reforming agenda: Brougham's Society for the Diffusion of Useful

Knowledge, its *Penny Magazine* and cut-price publications from Charles Knight, as well as the setting up of Mechanics' Institutes around the country, hoped to educate the working classes with a particularly middle-class conception of suitable information and skills.[12] The success of the Bridgewater Treatises, for example, a series of natural theological works on the latest research on scientific topics from geology to anatomy funded by the will of the Earl of Bridgewater, was just one signifier of widespread interest in the sciences, from the aristocracy to artisans.[13] Authors experimented with different literary genres in which to write scientific texts, increasingly marketed to a self-conscious 'reading public' for the sciences.[14] Indeed, the writing and reception of printed material proved a key arena for clarifying the role and remit of scientific practitioners, and the sciences themselves.[15] Responses to works such as the anonymous 1844 publication of *Vestiges of the Natural History of Creation*, a journalistic survey of the transmutatory 'universal law of gestation', from respected astronomy and geology to dubious claims over spontaneous generation of organisms, or the increasingly discredited claims of phrenology, forced the scientific community to insist on the pre-eminence of men of science as specialist practitioners, with hands-on knowledge of their subjects.[16]

By mid-century, scientific specialisms were firmly established: reified in separate societies and associated expert publications, these institutional trappings formed a base from which to lobby for greater state support.[17] Influential figures such as Thomas Henry Huxley and John Tyndall became significant figures in public life, arguing for the reform of educational systems, and the creation of teaching and research institutions that would match Continental rivals. The rate of production of scientific works for all audiences—but especially children—increased markedly, with a proliferation of periodical titles as well as songs and poems, novels, long-running book series, and school textbooks. Participation in local societies became one common means of engaging in scientific practice, as the construction of formal career paths led to paid positions on imperial projects of surveying, telegraphy, or of public health and education. Just as expert scientific practice rearranged itself into a range of different disciplines, from astrophysics to zoology, so too did new sciences of childhood dictate how children should learn. Of course, debates over what was most important and appropriate to include in books and stories for the young had been raging since the origins of children's literature itself.

The wand of reason

There are good reasons why it might be surprising to a twenty-first-century reader that reasoned scientific books are not easily distinguished from more imaginative or fantastical writings: commentators have been claiming such a separation for the past two hundred years. Fairy tales, they say, were chased away from childhood at the end of the eighteenth century as the demand grew for 'rational' tales, just as their folkloric equivalents were hounded from brook and dell by the industrial revolution and Victorian scientific and technological culture.[18] For example, writing in 1801, poet Lucy Aikin claimed that 'dragons and fairies, giants and witches, have vanished from our nurseries before the wand of reason'.[19] Though she converted 'reason' into a wand-wielding wizard, giving it the appearance of the very fantastical creatures it was supposedly banishing, her sadness over the replacement of fancy with fact was palpable. Just one year later, essayist Charles Lamb gave a more vicious and less resigned assessment of the current age, when he famously attacked his fellow writers, singling out poet and author Anna Barbauld for particular epistolary scrutiny:

> 'Goody Two-Shoes' is almost out of print. Mrs Barbauld's stuff has banished all the old classics of the nursery: & the Shopman at Newbery's hardly deign'd to reach them off an old exploded corner of a shelf, when Mary ask'd for them. Mrs B and Mrs Trimmer's nonsense lay in piles about. Knowledge, insignificant & vapid as Mrs B's books convey, it seems, must come to the child in the shape of knowledge ... Science has succeeded to Poetry no less in the little walks of Children than with Men.—Is there no possibility of averting this sore evil? Think what you would have been now, if instead of being fed with Tales and old wives fables in childhood, you had been crammed with Geography and Natural History? Damn them. I mean the cursed

Barbauld Crew, those Blights and Blasts of all that is Human in man and child.[20]

Lamb blamed the writers of introductory books for being too literal in their writings (knowledge 'must come to the child in the shape of knowledge'); again, he directly discussed their works as if they were the stories they had in this case banished not from the nursery but to a dusty corner of the bookshop. Writers such as Mrs Barbauld and Sarah Trimmer were peddling 'nonsense'. This is precisely what another author, Priscilla Wakefield, had termed the types of books Lamb lauded: in the preface to her 1816 *An Introduction to the Natural History and Classification of Insects in a Series of Familiar Letters* she rejoiced that the increased numbers of children's books being published had meant that 'Nonsense has given way to reason'. She went further, to claim that 'useful knowledge, under an agreeable form, has usurped the place of the Histories of Tom Thumb, and Woglog the Giant'.[21]

Wakefield's assessment chimed with those who championed 'useful knowledge' and the apotheosis of reason as the basis for rational education and civilization: she would have approved of Mr Gradgrind. Educational theories building upon John Locke's influential treatise that conceived of the child's mind as a blank slate, and new types of lesson that moved away from the traditional catechism to involve sensory engagement with the surrounding environment,

were used as the basis for practical school lessons, as well as
introductory children's books.[22] Writers such as Barbauld
and John Aiken, and Sarah Trimmer, deliberately attempted
to write moral tales that dealt with ethical questions such as
the care for others and for animals. New scientific subjects,
such as chemistry, or favoured pastimes that encouraged
children to interact with their surrounding world, such as
natural history or astronomy, were taught through both
dialogue and also more imaginative presentations: the miscellany *Evenings at Home*, for example, included a reincarna-
tory animal fable in the form of the 'Transmigrations of
Indur', as well as a chemical conversation on a cup of tea.[23]
Alongside these literary productions, new juvenile objects
were designed, made, marketed, and used for instructive
recreation in the last quarter of the eighteenth century,
including jigsaw puzzles, table games, and elementary scientific instruments.[24] The introduction to the Brothers
Grimm's first English translation bemoaned in 1823: 'Philosophy is made the companion of the nursery: we have
lisping chemists and leading-string mathematicians.'[25] It is
clear that scientific subjects were popular, and important,
bases for these new writings and artefacts; but the rhetoric
of replacement obscures the ways in which they were heavily reliant on—and were intended to develop—social,
moral, and imaginative skills, as much as sensory or reasoning powers.[26] Indeed, another source of attack was more
insidious: the claim that fairy tales were not just irrelevant

to an Enlightened age, but actually and actively dangerous. Several authors scrambled to denounce the fantastical and laud the reasonable. For instance, W. F. Sullivan's *The Young Liar!!!* of 1817, told how its hero, the appropriately named Wilfred Storey, had been fed a literary diet of fairy tales by his equally appropriately named nursemaid, Fibwell, and had become a liar. This had tragic consequences, when Storey died after a university duel, which had been brought about by his mendacious habits.[27]

More recent scholarship has dismantled these false oppositions, which have nevertheless been so influential in determining what has been deemed 'children's literature'. New, more sophisticated ways of understanding the historical contexts of publishing and reading in the late eighteenth to the mid-nineteenth centuries have been brought to bear on this material, and show a far more complex and interesting picture than these simple narratives of banishment or replacement or battle allow for. It is clear, for instance, that despite these writers' rhetorical posturing, fairy tales were popular and published right through the eighteenth and nineteenth centuries. New ways of combining these different literary traditions were also invented, for instance the 'moral fairy tales', which sought to imbue fanciful writings with appropriate messages for their juvenile audiences. In these ways, they were very like the moralistic parables and periodicals produced by figures such as Mrs Trimmer.[28] You could find fairyland in reasoned works; indeed, one eighteenth-century image went so far as to locate fairyland in the brain itself, with the land of courage in the heart, and that of dumplings in the stomach.[29] Such writings attempted to use the fairy tale as a way to write about the concerns of the age; to convert timeless tales into something appropriate for the eighteenth century. This was a practice that would really take off in the nineteenth century, as we will see.

A fairy tale age

A fairy queen sat on the throne of Victoria's Britain, and she presided over a fairy tale age. The nineteenth century witnessed an unprecedented interest in fairies and in their tales, as they were used as an enchanted mirror in which to reflect, question, and distort contemporary society.[30] It was, therefore, highly appropriate that Benjamin Disraeli, royal favourite, could refer to his monarch in Titania-esque terms.[31] Fairies could be found disporting themselves throughout the century on stage and page, in picture and print, from local haunts to global transports. There were myriad ways in which authors, painters, illustrators, advertisers, pantomime performers, singers, and more, captured this contemporary enthusiasm and engaged with fairyland and folklore; books, exhibitions, and images for children were one of the most significant.

Fairies themselves also changed in the nineteenth century, from the malevolent human-like sprites of the Renaissance to sources of Romantic inspiration and genius to the (usually) child-friendly, insect-like creatures we are familiar with from Disney films today: Puck became Tinkerbell. This metamorphosis can be traced most sharply in their appearance as a key subject of contemporary paintings.[32] The close connections between nature and fairyland were figured in numerous ways, as the existence—as well as the appearance—of fairies was challenged. Often fairies appeared as part of the miniaturized natural world, either as flowers, as butterflies, as grasshoppers themselves, or as their Tom Thumb companions; but they also appeared in more angelic guise, one manifestation of the Victorian fascination with the supernatural, acting—like spiritualistic mediums—between the quotidian world and higher realms. In these ways, the fairies were actively reimagined and reused throughout the nineteenth century, as authors pondered the role they play, had played, or should play, in the natural order of things. Anthropologists even made fairies the subject of scientific analysis, as 'fairyology' determined whether fairies should be part of natural history or part of supernatural lore: just one aspect of the revival of interest in folklore. Was there a tribe of fairy creatures somewhere out there waiting to be discovered, across the globe or in the fossil record? Were fairies some kind of folk memory of an extinct race? Some authors have argued that this interest in the fairies, and attempts to recapture them, depict them, or conserve them, was 'a protest against the strictly useful and material' that was supposedly the order of the day in the middle two quarters of the century; these 'attempts to reconnect the actual and the occult' would re-enchant the world, and demonstrate the dangers of utilitarianism without a tempering of imagination.[33] However, rather than the overtaking of fancy by factual analysis, a more complicated relationship can be traced.

It was not just fairies and fairyland that were of interest to Victorian audiences: from the early nineteenth century, new translations and collections of fairy tales and folklore were published in Britain to readers' acclaim. In 1823, a translation of the Brothers Grimm's collected tales was published by Edgar Taylor as German Popular Stories, credited with 'awakening' the interest in fairy tales 'for children and adults'.[34] Translations of tales by Hoffmann, Novalis, and of the Arabian Nights, to mention just a few, followed.[35] A whole host of new fairy tales were also written: for example, Danish author Hans Christian Andersen's tales arrived in Britain in 1846 in a translation by Mary Howitt as Wonderful Stories for Children, and, as Jack Zipes has argued, 'guaranteed the legitimacy of the literary fairy tale for middle-class audiences'.[36] The successors to what have been identified as the eighteenth century's 'moral fairy tales' highlighted the particular problems of an industrializing society, such as John Ruskin's *King of the Golden River* (1841): it has been argued that such stories were recruited as surreptitious means of socializing the young.[37] From brief poems in periodicals to the many-hued array of Andrew Lang's fairy books, the fairy tale became a staple of children's publishing, and would remain so ever since.[38]

Myriad new types of tale poured from the presses, combining the plots, tropes, or characters of traditional tales with a range of different topics, from the empire to Christianity to local regions, to medieval nostalgia, to the natural world. There is hardly a topic that interested Victorian audiences that was not recast in fairy tale mould; not least, the sciences. Reactions varied to these chimeric tales, with some bemoaning their often overt didacticism, where fancy had been sacrificed to content. In a well-known 1853 essay, Dickens deemed such didactic writings 'Frauds on the Fairies', taking to the pages of his periodical *Household Words* to denounce the practice of weighing down the fairy tale with overtly moralizing meanings: in particular, the temperance versions of *Cinderella*, *Hop o' my Thumb*, and *Jack and the Bean Stalk* written by his erstwhile colleague George Cruikshank.[39] However, as this book reveals, the fairy tales of science can be set apart from the didactic masses.

The fairy tales of science

Reviewing the latest two-volume English edition of the Brothers Grimm's *Household Tales* in 1853, the *Athenaeum*, a gentlemen's weekly periodical, reflected on the 'renewed love' for such 'old-fashioned tales' to be witnessed in this modern age. Far from being outstripped by the trappings of Victoria's Britain, they had, in fact, 'kept pace with the steam-engine and the electric telegraph', just some of the 'advances' which made the 'present age' 'more extraordinary' 'than any other'. The author pondered the 'singular fact' of this relationship, musing that perhaps it was for the best 'that this balance has been maintained'. The 'analytical tendency of Science, which views the Universe simply in its details, might lead us into a morbidly-exclusive perception of the mechanical anatomy of things, were it not for Imagination, which feels and enjoys results by means of the instincts of the heart'.[40] However, many other ways of balancing, noticing, and commenting on the relationship between the sciences and the imagination can be found in the nineteenth century, particularly in what, echoing Tennyson, we can term 'the fairy tales of science'. The *Athenaeum* author continued, to note that 'Coincident with the world of Fact, in nearly all ages and among all nations, and lying by the side of that world like a fantastic shadow, has been the world of fairy Fiction'.[41] This 'fantastic shadow' may have been 'coincident with the world of Fact'; but, as I will argue, it was no coincidence.

Scientific engagement with fairyland was widespread, and not just as an attractive means of packaging new facts for Victorian children.[42] Some authors, like Margaret Gatty, confessed to a direct inspiration from the new fairy tales, having written her first stories 'in an outburst of excessive admiration of Hans Andersen's Fairy Tales'; moreover, the inspiration was to go beyond the fanciful story-telling to simultaneously drive home a deeper meaning: she regretted the fact that 'although he had, in several cases, shewn his power of drawing admirable morals from his exquisite peeps into nature, he had so often left his charming stories without an object or moral at all'.[43] The fairy tales of science had an important role to play in conceiving of new scientific disciplines; in celebrating new discoveries; in criticizing lofty ambitions; in inculcating habits of mind and body; in inspiring wonder; in positing future directions; and in the consideration of what the sciences were, and should be. A close reading of these tales provides a more sophisticated understanding of the content and status of the Victorian sciences: they give insights into what these new scientific disciplines were trying to do; how they were trying to cement a certain place in the world; and how they hoped to recruit and train new participants.

These often quirky, usually charming, and occasionally dull stories were an important new way in which nineteenth-century Britons enthused about, communicated, and criticized the sciences. From nursery classics such as *The Water-Babies* to the little-known *Wonderland of Evolution*, or the story of insect lecturer *Fairy Know-a-Bit*, the fairies and their tales were often chosen as an appropriate new form for capturing and presenting scientific and technological knowledge to young audiences. Fairies and imps, dragons and demons, giants and gnomes, appeared throughout these texts: as framing devices, as storytellers, as starring characters, as illustrations, as the invisible forces of nature. They demonstrate how, for many, the sciences came to replace the lore of old as the most significant source of marvel and wonder, and of fairy tales themselves. Far from being the destroyer of supernatural stories about the world, through these fairy tales the sciences were presented as being the best way to understand both contemporary society and the invisible recesses of nature, since they revealed the hidden magic of both the sciences and of everyday life. But they also claimed a higher status: that their enchantment revealed the true wonders of nature. These real fairy tales of science were avowedly stranger than fiction. The history of the Earth, forces of the universe, inhabitants of a beehive, or contents of a drop of water, contained more wonderful characters and magical transformations than the most exotic lore of old.

We first turn to the brand-new geological sciences of palaeontology and mineralogy, which questioned the history and composition of the earth, and introduced its previous inhabitants to astounded audiences. We consider the new relationship Victorians had with time, and how this precise delineation of past eras had consequences for the once-upon-a-time of fairy tale fame. We will investigate how this affected and was affected by previous stories of monstrous creatures roaming the countryside: the impact on the relative status of folklore and these new scientific tales, as dinosaurs were pronounced superior to dragons. Rabbit-holes and looking-glasses may have their place, but the surest way to wonderland was to be found in practising the sciences.

1

Once Upon a Time

'there is, after all, an unreality about themselves...'
[John Lindley], on the 'Saurian monsters in Penge Wood'
Athenaeum (1854)[44]

Once upon a time, monstrous beings roamed the earth. Soaring, lumbering, pouncing, snarling, these fearsome creatures would terrorize the countryside, devouring flocks of livestock, taking princesses captive, and even fighting knights. Dragons and griffins, it was said, were common sights in the historic landscape, as they left their rocky lairs to prey on the people with fiery breath. By the mid-nineteenth century, thankfully, this was a vanished era: merely legends lived on from that mythic age. But some argued that it was only in recent years that the true sources for these legends had been found: retrieved remnants of previous worlds hidden in Dorset cliffs and in Yorkshire caves. The

remains of these beings were now being systematically procured, analysed, classified, drawn, reconstructed, and displayed, as part of a new science, called geology.

Geology, we will see, had a particularly close relationship
with fairy tales, with identifications often being made
between fragmented fossil discoveries and creatures of
lore. The basic ideas of what has become known as 'deep
time' were formulated directly in association with legends of
giants and monsters, as the exhilarating discoveries of palaeontology and mineralogy claimed to tell what had really
happened once upon a time. Authors such as John Cargill
Brough, in his Christmas book for boys that showcased *The
Fairy-Tales of Science* (1859), argued that palaeontology resurrected an 'Age of Monsters', replacing 'shadowy griffins and
dragons' with 'real and tangible' beings.[45] These new objects
of scientific investigation, authors such as the Reverend
Hutchinson showed in his *Extinct Monsters* (1892), were
allegedly the actual creatures of legend.[46] But this easy
conflation meant that geology's artefacts of study and
methods of analysis were most at risk of being dismissed
as imaginative concoctions, fevered dreams of wishful nat-
uralists who populated deserted ancient landscapes with
ferocious beasts. Therefore, the superior status of geologists'
investigations was constantly affirmed, not only in terms of
their deepening of scientific understanding but also, crucially, in terms of their storytelling potential. The so-called
'antediluvian monsters', these authors argued, provided even

more fantastical beasts and gruesome giants, marvellous processes and fabulous histories; and, as the often bloodthirsty visual depictions made clear, could be even more terrifying. Who needed dragons when the real stories of monstrous creatures and deadly battles were now revealed for all?

The age of monsters

John Cargill Brough, chemist and journalist, lived in the age of monsters. This was both the title he gave to the opening chapter of his 1859 book, and also an apt description of Victorian Britain, where wide audiences thrilled to news of palaeontological discoveries being made in caves and cuttings and quarries around the country.[47] The book's title and epigraph were explicitly drawn from Alfred Tennyson's 'Locksley Hall', a meditation on past, present, and future, which in many ways summarized the consequences of seaside fossil-hunting.[48] Newly conceived of as the fossilized remains of previously existing creatures, anything from ammonites and coprolites to entire preserved reptilian skeletons could indeed be encountered by children wandering Britain's beaches. From the 1830s onwards, telling the stories of these strange creatures, explaining what they were like, how and when they lived and behaved, how they had died and survived, had become a popular topic for children's books, with geological conversations, textbooks, meditations, and even autobiographies and fairy tales emerging from London's printing presses.[49] Though the Laureate's telling phrase might have suggested otherwise, *The Fairy-Tales of Science* was not in fact a set of magical stories about these new discoveries. Rather, it was a series of succinct summaries of past and recent work, organized by scientific subject, which endeavoured to introduce a broad range of relevant topics and knowledge to childish minds, including very similar information to contemporaneous publications which were issued in alternative generic guise. For Brough, this choice of literary form would 'divest' the novel scientific information his book contained of 'hard and dry' technicalities, reclothing it 'in the more attractive garb of fairy tales'. He confessed this was 'a task by no means easy'.[50] Indeed, some could argue that Brough's introductions read very like the standard dry presentations he had wished to avoid; enlivened solely by the fabulous accompanying illustrations by Charles H. Bennett.[51] But it is clear that Brough intended and achieved much more than simply dressing-up the sciences in an 'Emperor's new clothes' of fanciful prose, as his choice to begin the collection with geology made clear.

In 'The Age of Monsters', the fairy tale element of Brough's chapter was to be found nested within his framing discussion of the heroic modern men of science: underneath its clothes, in fact, rather than on its surface. Brough's prose traced exactly the kind of progressive narrative from rude

superstition to modern understanding that he wished his
youthful readers to undergo. Starting out with a slightly
absurd sketch of how previous ages had hymned the
power and predation of the dragons—the infantile nature
of such tales reinforced with his choice of nursery-rhyme
resonances ('All the King's horses and all the King's men
were powerless in the face of such a foe')—Brough set up
the Victorian Briton as a decidedly superior sort. 'Banishing'
the knightly exploits of Saints George and Patrick to 'credu-
lous times' in which the 'monsters of enchantment' were
believed in, he traced the more recent and more heroic
example of the 'curious men' who had instead 'ventured to
pry into the secrets of those terrible caves'.[52] Throughout
the chapter, a contrast was drawn between the superiority
of the nineteenth-century outlook and previous ages of
civilization; characteristic of the Victorians' renewed interest
in previous ages, as so apparent in Pre-Raphaelite art, Gothic
architecture, or Arthurian poetry. However, the twin ages of
the monsters were also connected, for instance with a direct
comparison between two different kinds of monsters, these
beasts and the gigantic iron engineering projects of Isambard Kingdom Brunel, of Leviathan (Great Eastern) fame. Other
mediating techniques were put to work to facilitate the
passage's voyage back in time, providing more familiar
stopping-off points; for instance, the almost Homeric
catalogue of ships that got progressively older: 'no iron
steam-ship, no vessel of war, no rude canoe even'.[53] Other

contemporary books also used the classical world as a stepping-stone to the past, perhaps most prominently the Mediterranean ruins which graced the cover of Charles Lyell's influential *Principles of Geology* (1830–3).[54]

But Brough gave his readers no such warning as to their sudden dislocation in time, immediately plunging both his narrator and reader into the past at the beginning of a paragraph, with a startling switch to the first person plural, and the present tense. Suddenly, 'We are on the shores of the Ancient Ocean.'[55] Brough stressed the alien nature of this past world to Victorian eyes, reaffirming the expert work that had to go into conjuring it for his readers—this was no task for the unskilled speculator.[56] At first, narrator and reader 'search in vain for any sign of Man's handiwork', and are frustrated by their attempts to comprehend the novel beings and vistas before them. The 'teeming' ocean 'contains no single creature which has its exact likeness in modern seas'; the 'savage panorama' is 'utterly unlike that of any modern land', and kindles 'mingled feelings of admiration and awe'.[57] An overall appreciation of the 'wonders' with which they are 'surrounded' is all that can be felt, before particular objects are noticed. This generalized sense of wonder and curiosity was often deemed the essential starting-point for further scientific investigation: the fairy tale framing that Brough employs is able to emphasize this much more readily than less narrative works.[58] Just as with the comparisons to the more familiar classical or mythic worlds Brough's narrator therefore must deploy a series of comparisons in order to communicate something of these creatures' appearance, scale, and behaviours. Familiar references are layered in a sophisticated way to achieve this aim: for instance, the first creature encountered is introduced with an allusion to folklore and fairy tales that have previously been discussed in the chapter: 'Yonder is one of these old monsters of the deep.'[59] Keeping the visual directions ('Yonder') at the forefront of the narrator's sentences encouraged the reader to envision this mental hybrid, which would—like the process of palaeontological analysis and comparative anatomy itself—be assembled from parts and analogies. Next, a footnote introduced both the unfamiliar name for this creature ('Cetiosaurus'), but also its etymological translation, which gave a clue as to its appearance and habits, and also its strangeness. Combining two modern creatures, the *Whale-like Lizard* was not a whale, but it is maritime, and hence whale-like. It is also a lizard; but evidently not a lizard that the reader is likely to have seen, 'whale-like' not being a common adjective for the reptile family. Then, the narrator gave a further idea as to the particular aspect in which it may be whale-like: its scale, since it 'might easily be mistaken for some rocky islet'. In these ways, Brough's prose played on the word 'rocky', hinting back to what now remains of this creature, and also how the creature's appearance had been put together, rather differently though in an analogous way to this one here, through the power of comparative anatomy.

Comparative anatomy, rather than fictional device or literary allusion, was the more usual process by which nineteenth-century palaeontology travelled through time. Like Brough's introduction to the Cetiosaurus, it used comparisons to living creatures to resurrect their long-extinct analogues. Its most famous proponent was Georges Cuvier, working at the Paris Museum of Natural History at the turn of the century.[60] An established science by the time Brough was writing his book, palaeontology nevertheless could be charged as imaginative and potentially dangerously speculative: just like fiction writers palaeontologists were really inventing monsters from thin air, and imposing plots onto the history of the world, casting themselves as heroes in such narratives.[61] Brough highlighted the palaeontological process towards the end of his chapter on 'The Age of Monsters', as he took the opportunity to introduce the practice of the men of science. He gave his reading interlocutor a sceptical voice, as he posed a series of rhetorical questions that challenged the veracity of the sights encountered on the literary trip through time. 'How', the reader was invited to ask, 'can we know anything' at all about the past? 'How can we come to any conclusion'? Gratifying his reader with the acknowledgement that these questions 'will probably occur to the reader', he used the labours of the geologist to combat any 'doubts' as to the reality of the foregoing narrative. Indeed, the truthfulness of what has been depicted was affirmed through a direct connection with nature, gained through the privileged insights of the geologist. Again drawing on a comparison to a process and an object of which his readers might be aware, he detailed how the 'monsters have been their own historians', and have 'described themselves' in the 'Stone Book' of the landscape. To further cement the literary analogy, this choice of metaphor referred back to the image with which the book opened, one of Charles H. Bennett's monsters writing in a stony book; it evoked a particular kind of book—an illustrated volume like the one the reader held in his or her hand; and also drew attention to the flat fossil remains that could look like ferns, flowers, or insects pressed within the pages of a book. Brough listed the myriad types of evidence that had survived to the modern day, and that were studied by geologists—very different from the lively creatures the reader had encountered on the trip to the past: 'broken and water-worn bones, detached teeth, fresh-water shells, fragments of trees, and even the bodies of insects'.[62] Indeed, Brough was at pains to emphasize what hard work it was to resurrect these creatures: it was no mere fancy of the imagination, rather an imperial feat worthy of the age. Only through the 'untiring industry and perseverance' of the geologist could those bones and teeth and shells and trees be assembled, analysed, and—at last—the man of science would be 'rewarded with a vision'.[63] Crucially,

Brough stressed the 'industry', the patient labour, the work that had gone into this process. Unlike other ways in which such visions were presented—which, as we will see, included magical carpets and fairy guides—here it was hands-on palaeontological work that was the master key.

Dragonology

With his opening chapter of his book, Brough introduced his juvenile readers both to the age of monsters as it had been but also to the age of monsters in which they lived (see Fig. 2). Through literary techniques that transported his readers back in time (with a helpful narrative voice to guide them), and through familiar comparisons, children could learn about the most famous new monsters of the day, from the sea-monsters to the Megalosaurus. Moreover, through his evocation of the heroic business of the gentlemen of science and what one had to do in order to bring these creatures back to life, Brough revealed the current enthusiasm for these discoveries. But for some, verbal descriptions and simple illustrations were not enough to bring back extinct creatures: publishers and audiences turned instead to the world of art and to more sophisticated images to capture the sublimity of the past. The kinds of creatures and battles that Brough had described in print were brought before the eyes of audiences by the pencil of

Fig. 2. 'The Age of Monsters', frontispiece to John Cargill Brough, *The Fairy-Tales of Science* (1859). In this image by Charles H. Bennett, a prehistoric creature inscribes its own entry in the fossil record, which is depicted as a stony book. The representation of a snarling animal below emphasized the presentation of such vanished beings as the true monsters of legend.

antediluvian artists. And what they often depicted were dragons.[64]

Brough stressed that the kinds of creatures called up by this reasoned process based on actual evidence provided superior fodder to mere fancies of the imagination. For instance, when the reader and narrator encountered a Plesiosaurus, comparison (as with the introduction of the Cetiosaurus) was made to 'a fish-like body, a long serpentine neck, and the tapering tail of a lizard'; yet this combination of attributes surpassed all of these creatures. 'The imagination of man', Brough wrote, 'never called up a shape so weird and fantastic as this.'[65] Modern science, he argued, conjured a crazier creature than any chimera from the tales of old. Not just in terms of appearance, but also in terms of behaviour was this modern monster more than a match for its folkloric predecessors:

> Look at him now, how eagerly he pounces upon every living thing that comes within the range of his pliant neck, how cruelly he crushes the bones of his victims, and how greedily he swallows them! We never witnessed such unhandsome conduct in a monster before. Leaving him at his disgusting banquet...[66]

With its echoes of Brough's earlier introductions to fairy-tales of dragons and griffins, this section deliberately set up the Plesiosaurus as a being far from pleasant to encounter, and instead as something akin to an ogre, giant, or other monstrous creature from the fairy tale cast of characters. Verbs such as 'pounce', 'crush', or 'swallow' brought to mind stomping ogres and gluttonous giants: even the reference to the 'disgusting banquet' evoked a medieval feast disturbed by an evil creature. It is quite clear from Brough's prose that this creature surpassed all previous ones in its monstrosity ('never witnessed...before'); furthermore, this was a real creature that has been resurrected through a scientific process, which made its behaviour all the more horrifying.

The set-piece battle of 'The Age of Monsters' had also been a fight between two 'dragons', encountered by the narrator and his reader-companion. Lively prose with vicious verbs, lots of action and noise as well as sights, helped frame this as a duel. Whereas a fanciful contemporary image by artist Benjamin Waterhouse Hawkins had reworked the legend of St George and the Dragon to depict St George and the Pterodactyl, in Brough's story no saintly hero was at hand:

> But hark! What noise was that? Surely that harsh discordant roar must have proceeded from the deep throat of some monster concealed in yonder forest...Now a crashing among the trees, followed by a wild unearthly shriek.
>
> Look at that terrible form which has just emerged from the thicket. It rushes towards us, trampling down the tall shrubs that impede its progress as though they were but so many blades of grass....Its head is hideously ugly, its immense jaws and flat forehead recalling the features of those grim monsters which figure in our story-books. Its dragon-like appearance is still further increased by a ridge of large triangular bones or spines which extends along its back. (*The Hylaeosaurus, or *Wealden Lizard.*) We should not be at all surprised were we to see streams of fire issuing from the mouth of this creature, and we look towards the palm-forest half expecting a St. George to ride forth on his milk-white charger.
>
> See!—some magic power causes the trees to bend and fall— the dragon-slayer is approaching! Gracious powers! It is not St. George, but another Dragon nearly double the size of the

first. He proclaims his arrival by a loud roar of defiance...The newcomer is certainly a very sinister-looking beast. His magnitude is perfectly astounding...(*The Megalosaurus, or *Great Lizard*)...[67]

Footnotes gave—as with the other creatures Brough introduced—both the correct scientific and etymological name for the prehistoric beast he was describing, but otherwise the prose could have described any other meeting between two monstrous forms. This manner of description continued throughout the anecdote, which gave the reader a ringside seat to the combative, onomatopoeic encounter:

> Alas! there is no escape for you, unfortunate Dragon!...Now they meet in the hollow with a fearful crash...What a fearful conflict! How they snort and roar!...The hero of the crest is the first to rise—he makes off towards the forest, and may yet escape. Alas! He falls exhausted, and the great monster is on his track...He approaches his fallen enemy. Now he jumps upon him with a crushing force, and now his enormous jaws close upon the neck of his victim, who expires with a shriek of pain.[68]
>
> We can gaze no longer at this awful scene. The battle was sufficiently exciting to absorb our attention...[69]

Heroes and enemies, exciting battles and shrieks of pain: here was action enough for any medieval saga or classical epic.

Such engrossing descriptions were only one way in which audiences could experience the 'awful scenes' from deep time: artists also were put to work in imagining and presenting these vanished worlds, as anything from small lithographs to privately circulated caricatures, to public panoramas.[70] The most famous geological artist was John Martin, a painter of biblical epic, master of chiaroscuro, and inventor of the 'Martinesque' style of prehistoric illustration that was to prove canonical.[71] His famous frontispiece for Gideon Mantell's *Wonders of Geology* (1838) was widely imitated in other introductory books, for instance in the frontispiece to John Mill's imaginative children's book, *The Fossil Spirit: A Boy's Dream of Geology* (1854).[72] Like Brough, Martin chose to represent a moment of conflict, a battle between fearsome creatures, but he could also introduce the primeval landscape beyond.

Just as Brough had used the dragon as a point of reference for his literary depiction, so too were traditions of drawing dragons exploited by antediluvian artists. Such a visual continuity helped reinforce the close relationship—indeed, identification—between dragons and the new prehistoric creatures in the nineteenth century.[73] The dragon was both claimed *as* reptilian and used as a more familiar reptilian model, fossil discoveries being used to argue both for the reality of the more familiar mythic creature and also for its redundancy. Indeed, it was only in the recent past that the dragon had been conceived of as a potentially real but disappeared creature: even in Samuel Johnson's 1755 *Dictionary* they were only described as 'perhaps imaginary'.[74] Artists in particular played on these similarities when seeking visual referents for their depictions: hence, though Brough claimed that the 'monsters of romance were nowhere to be found' ('Triumphant science had banished them from the realms of fact' and the 'poor ill-used Dragon has now no place to lay his scaly head, the Griffin has become a denless wanderer, and the Fiery Serpent has been forced to emigrate to a more genial clime!'[75]), they were in fact to be found in palaeontological imagery all around.

This association was more than just a loose association between large scaly creatures, however: and identifying it allows us to glimpse the crucial relationship between folklore and the sciences as both shifted radically in scope and structures across the nineteenth century.[76] Some figures, such as Scottish geologist Hugh Miller, 'walked between two worlds' of folklore and scientific expertise, and close connections have been traced between his working-class Cromarty background, his practical skills with stone as a mason, and his geological work on, for instance, the Old Red Sandstone.[77] Other figures such as Robert Hunt participated in the burgeoning field of folklore studies as well as working in scientific institutions, and writing scientific works in a range of genres: collecting *Popular Romances of the West of England* from his Cornish and Devonshire homeland could be set alongside reviewing for leading periodicals, giving introductory lectures, conducting experiments in the new science of photography, or even experimenting with new forms of scientifically accurate fantastical fiction.[78] Just as these individuals proved multifaceted, so too were the debates: for nineteenth-century audiences as well as practitioners these questions and answers and practices and stories were interestingly intertwined. It was not so far a leap from fairy tales to science, or from science to fairytales. The Brothers Grimm were linguists, after all. Indeed, by the 1890s and the publication of Hutchinson's *Extinct Monsters*, dragon comparisons were deemed the best way to introduce his subject-matter to new audiences. As he began, echoing the kinds of claims that Brough had made decades earlier:

LET us see if we can get some glimpses of the primaeval inhabitants of the world ... We shall, perhaps, find this antique world quite as strange as the fairy-land of Grimm or Lewis Carroll. True, it was not inhabited by 'slithy toves' or 'jabberwocks', but by real beasts, of whose shapes, sizes, and habits much is already known—a good deal more than might at first be supposed. And yet, real as it all is, this antique world—this panorama of scenes that have for ever passed away—is a veritable fairy-land. In those days of which geologists tell us, the principal parts were played, not by kings and queens, but by creatures many of which were very unlike those we see around us now. And yet it is no fairy-land after all, where impossible things happen, and where impossible dragons figure largely; but only the same old world in which you and I were born. Everything you will see here is quite true. All these monsters once lived. Truth is stranger than fiction; and perhaps we shall enjoy our visit to this fairy-land all the more for that reason. For not even the dragons supposed to have been slain by armed knights in old times, when people gave ear to any tale, however extravagant, could equal in size or strength the real dragons we shall presently meet with, whose actual bones may be seen in the Natural History Museum at South Kensington.[79]

As Hutchinson intimated, the three-dimensional remains of extinct creatures to be found exhibited in London were superior to any dragon of old. Visitors to another popular metropolitan destination, the Crystal Palace, would have agreed.

Fairyland in 'fifty four

In December 1853, *Household Words* knew who the modern genie was, hailing Joseph Paxton as 'Djin' of Sydenham.[80] This pioneering conservatory-builder had been the creative force behind the so-called 'Crystal Palace', an affectionate moniker which could have come straight out of an Arabian fairy tale. But this was no fantastical building: the transparent marvel and palatial surroundings were instead wonders of modern science and technology (see Fig. 3). The glass structure had first been erected in Hyde Park in 1851 as part of the Great Exhibition of the Works and Industry of all Nations. Three years later, newly expanded, the Crystal Palace reopened in East London as a commercial enterprise.[81] But the building and its contents had not only expanded in size: they extended even further back in time, as its internal courts were joined by a sequence of twenty-one antediluvian creatures resurrected in its grounds.[82] Audiences could now go beyond the literary or visual

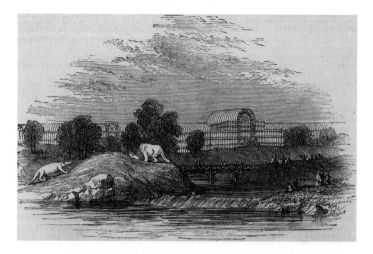

Fig. 3. The Sydenham Crystal Palace, as seen from the geological islands, with several prehistoric animals visible in the foreground, from P. H. Delamotte, 'The Geological Island', in Samuel Phillips, *Guide to the Crystal Palace and Park* (1854). Lurking on islands in the lake at some distance from the Palace, the animals seemed at a temporal as well as spatial remove from the glass marvel of Victoria's Britain.

presentation of prehistoric creatures, to witness their three-dimensional reconstruction.[83]

Charles Dickens's periodical, *Household Words*, which in 1854 was serializing his novel *Hard Times* that made so much of the contested relationship between facts and fancy, had the previous December taken a trip to Sydenham.[84] Just as *Hard Times* saw in the industrial landscape of the north an 'elephant in a state of melancholy madness' rather than the piston of a steam engine, for the article's authors, George Augustus Sala and W. H. Wills, the new development of

the People's Palace was, instead, to be viewed as 'Fairyland'.[85]
It was characteristic of Household Words to bring together
imaginative strategies of writing with introductions to issues
of the present day: for instance, another article titled 'Our
Phantom Ship on an Antediluvian Cruise' combined geological exposition with a ghostly voyage on the seas of time,
and the tropes of travel narratives.[86] It was commonplace in
the mid-century periodical press to run samples of different
literary genres cheek by jowl on neighbouring print columns; but these Household Words articles went one further, to
play around with means of clothing information within
their paragraphs themselves.

'Fairyland in 'Fifty Four' immediately invoked the fairy-
tale muses, as the article's narrator implored them to rescue
him from all of the most recent scientific institutions and
celebrities. Drawing on the familiar and much-abused trope
of the 'death' of fairy lore due to the progress of modern
society, the narrator gave a rather indiscriminate catalogue
of fairy tale and magical authorities, immediately compared
with a scientific equivalent:

> O, BROTHERS GRIMM; O, Madame D'Anois, O, Sultana Sche-
> herazade and Princess Codadad, why did you die? O, Merlin,
> Albertus Magnus, Friar Bacon, Nostradamus, Doctor Dee, why
> did I implicitly believe in your magic, and then have my
> confidence utterly abused by Davy, Brewster, Liebig, Faraday,
> Lord Brougham and Dr. Bachhoffner of the Polytechnic Insti-
> tution? What have I done that all the gold and jewels and
> flowers of Fairyland should have been ground in a base

mechanical mill and kneaded by you—ruthless unimaginative philosophers—into Household Bread of Useful Knowledge administered to me in tough slices at lectures and forced down my throat by convincing experiments?[87]

Something of Dickens's critique in *Hard Times* seeped into this analysis, particularly the echo of Gradgrind in the 'base mechanical mill', which 'ground' gold into dust. The narrator charged the men of science with a lack of imagination, and bemoaned their propensity for an unpalatable diet of 'convincing experiments', which they were determined should be fed to all. Such responses were a common part of the debate over who should have access to scientific knowledge in the second quarter of the nineteenth century, and were references that this middle-class periodical's readers would have picked up on: for instance, the allusion to the Society for the Diffusion of Useful Knowledge in the 'Household Bread of Useful Knowledge', and to its driving force, Lord Brougham, as well as to those stars of the Royal Institution, Humphry Davy and Michael Faraday.[88] But this was, as the reader suspected all along from the plaintive comic tone in which it was written—not to mention the periodical's usual enthusiasm for scientific and industrial knowledge—merely a wordsmith's ploy; an extended rhetorical question to which Sydenham was then introduced as the answer. Joseph Paxton, identified with Chatsworth and the *Victoria regia* waterlily—itself named after Disraeli's fairy queen—became instead a 'burly Djin in a white hat and a frock coat with a huge lily in the button-hole', master of the crystal:

> 'I will conjure for you, by the aid of my crystal (a million times bigger and clearer than the crystal of Raphael the astrologer), a fairy palace with fairy terraces, and fairy gardens, and fairy fountains, compared to which the palace of Sardanapalus was a hovel, and the gardens of the Hesperides a howling waste. You shall see, through my crystal, so far into the past, that the retrospection shall not end until the world before the flood is revealed to you, with the fat, slimy, scaly monsters which then had life upon it...'[89]

The narrator stressed the superiority of the modern marvel to those of old, as had John Cargill Brough in his introduction to the 'age of monsters': just as the need for dragons had been superseded by the dragon-like hylaeosaurs, iguanodons, and megalosaurs, so too did these places stand in comparison as like palaces to hovels, or gardens to wasteland. Crucially, the superior scientific vision provided deeper and further 'retrospection' than ever before: 'so far into the past' that 'fat, slimy, scaly monsters' were revealed, older than (and, it is implied, therefore superior to) medieval myth.

If the article's message lay in comparison, its humour rested on conflation, in particular that mapping the precisely evocative trappings of Victorian Britain onto the more timeless tropes of the fairy tale. The policemen who guard the entrance 'to Fairyland in Crystal', for example, were described as 'blue, and uncompromising' 'gnomes': to 'judge from the costumes of these gnomes you would take them to be plain constables of the Metropolitan Police; but, take my word for it, they have all the gnomical etceteras beneath their uniform and oilskin'.[90] The cod-scientific tone of 'gnomical etceteras' once again highlighted the expert knowledge appealed to, as well as the readers' assumed knowledge of what those 'etceteras' might be, as their previous experience of fairy tales helped to fill the Latinate lacuna. Throughout, the article played on the comic juxtaposition of what would be the expected inhabitants and plot devices of fairy tales, and the more mundane actions of visiting a suburban London park. For instance, 'the entrance to Fairyland is not effected by rubbing a lamp, or clapping the hands three times, or by exclaiming "Open Sesame"': more boringly, it is by paying a fee.[91] The bathos of the piece ran throughout, emphasizing, for example, the 'tools and fragments of planking and old ropes' with which the 'hall of the Fairy Palace is strangely strewn': these kinds of everyday domestic details were missing from many traditional tales, and also highlighted the industry underpinning the spectacle.[92] Workmen themselves appeared, but recast and reclad as 'fustian fairies' going about their business; a far cry from either the poor orphans or the captive princesses more usually found as fairy tale protagonists. But it was when the narrator stepped outside of the Palace to 'traverse the parks and gardens of Fairyland' that the bathos became most acute, for the narrator immediately encountered 'that extremely mundane attribute, mud'.[93]

The muddy encounter and succeeding wordplay immediately foregrounded the geological basis for the monsters'
re-creation: they had been built from the mud (rocks underfoot, muddy geological digs); they had been sculpted with
mud (clay); and they could teach audiences a lesson about
resurrection (echoing the biblical teachings with which they
would be familiar). The precise—if muddily precarious—
walk down through Sydenham's grounds traced in reverse a
trip back in time: 'through a wood, and across several planks
over gulleys, and through many morasses, quagmires, cart-
ruts and ditches...at last [to] arrive at a long low shed...
and finally the world before the flood'.[94] The journey
through time was appended to the journey through space
as though it was not of another order at all, a message
reinforced by the way in which the courts and grounds at
Sydenham were laid out.[95] To get to 'the world before the
flood', as we encountered it in Brough's book, was not a
question of faith or of an imaginative leap; it took diligent
work, it took planks and ditches, and men of science and of
art. In this case, that 'long low shed', a staging post on the
way to the past, was revealed as the workshop of sculptor
Benjamin Waterhouse Hawkins, 'an astute Triton in Hessian
boots', who, with the aid of palaeontologist Richard Owen,
'the King of Animals', had been responsible for re-creating
the creatures (see Fig. 4).[96] Or, as the narrator of the article
had it, bringing 'back those antediluvian days when there
were giants in the land'.[97] The creatures were—like their

Fig. 4. 'The Extinct Animals Model-Room at the Crystal Palace Sydenham', from *The Illustrated London News* (31 December 1853). This image from a report into the making of the 'antediluvian animals' took readers of the *Illustrated London News* inside Benjamin Waterhouse Hawkins's crowded workshop on the Sydenham grounds. Glimpses of several of the models can be seen, as well as tools of the sculptor's trade, and a couple of living animal inhabitants.

literary counterparts—compared to other living analogues ('gigantic creatures of lizard, toadlike, froglike, beastlike form'), who would 'grin at you, crawl at you, wind their hideous tails round you'; but the process of 'scientific art' by which they had been produced was also foregrounded:

> All of which is explained to us in a little studio, where sepia sketches of elks and mastodon, and megatheria mingle with clay sketch models and casts of skulls and femurs of fossil mammalia and reptiles.[98]

The end of the article mirrored its opening, with a rhetorical plea to the reader ('do you think I shall have been guilty of exaggeration in calling it Fairyland?').[99] This time, rather than asserting how modern sciences and industry had killed off the fairy tale, the narrator made a slightly different case; rather, that 'magic and magicians are not dead', as scientific and inventive figures and their accomplishments were in fact their latest incarnation, a quite literal incarnation in the case of the monster models.[100]

Reviewing a selection of introductory scientific books a year after the display opened, Margaret Oliphant put paper productions aside for a paragraph to reveal her thoughts on the Sydenham display. Unlike the authors of the *Household Words* article, she was not impressed by the geological, technological, and sculptural feat of 'scientific art' to bring the antediluvian creatures back to three-dimensional life. Rather, she had preferred to let her own imagination loose on the skilful prose of writers such as Hugh Miller, to paint her own pictures of the creatures. Now, 'Professor Owen's "restorations", however true they may be, are rather a damp upon the fervour of geological visions'.[101] Instead of the glorious panoramas she had been wont to conjure in her mind's eye, the concrete reality before her presented a rather different sight:

> But, heaven help us, what are these?—these frightful scaly monsters—these giant reptiles—these gaping jaws, and eyes in which no speculation dwells? Are these the heroes of our earliest romances?...it is rather hard upon an author to take the poetry out of him after this remorseless fashion. Let science have her will of her own gigantic offspring; but poetry, we are afraid, cannot look a second time into these fishy eyes. Inexorable fact and Professor Owen have made an end of all our pretty pictures.[102]

Whereas Hawkins, for instance, saw in his project of visual education an opportunity to present audiences with 'the things themselves', thus removing the work that would have had to go into envisioning the creatures for themselves, for Oliphant this removed, also, the pleasure of the experience, rendering it instead an unwelcome confrontation with the 'fishy eyes' of the past. This was a particular problem for young visitors: she went on to address her childish readers in particular, imploring them to avoid the creatures wherever possible: 'we beg of every young geologist...to close his eyes very hard as he comes towards the fairy palace, and never, for any inducement, to be tempted to stray far into the grounds.'[103] What had been, for Oliphant, a delight in prose could have different consequences when cast in stone: the uncanny creatures disturbed. Therefore, although in writings about these creatures we can see again an emphasis on the work that had to go into creating prehistoric vistas, such labour existed in tension with the projects of visual education espoused by figures like Hawkins, who wished to elicit precisely the kinds of direct, visual, and visceral experience of these creatures captured by Oliphant's review. It was now too easy for imaginative leaps to be made.

What was in other ways a sign of monstrous superiority—the 'terrible' in the 'terrible lizards'—was here a disadvantage, scaring children far more than the trolls and ogres of old. Of course, any adult reading these words in *Blackwood's Edinburgh Magazine*, an eminently respectable periodical, might have had trouble convincing their offspring not to go and look at the monsters, and Oliphant's tone here suggests she might not be entirely serious: surely part of the appeal of the display was in its very monstrosity, just as Brough's stories played up the violence and combat of the antediluvian world. Some children did, apparently, find the display disturbing: *Punch* magazine suggested one possible reaction of contemporary children, depicting 'Master Tom' 'strongly objecting' to having 'his mind improved' by a visit to the antediluvian display.[104] Dwarfing the poor child, the fearsome beasts seemed to crowd around the be-hatted boy, and the placid refuge of the country spire seemed a long way off in the distance. These particular monsters, then, became part of a more generalized conversation ongoing at least since Locke's 1693 *Some Thoughts Concerning Education*, when he warned that '*Sprites* and *Goblins*' could 'sink deep' into children's minds.[105] Late eighteenth-century authors cautioned against the 'terrific images' that could be produced in juvenile imaginations by exposure to these kinds of fantastical creatures, and wrote stories such as 'The History of Francis Fearful' to preach this doctrine.[106] A similar fate to that of Francis Fearful was depicted again by *Punch* in George Du Maurier's image 'A Little Christmas Dream': the illustrator claimed to have replaced his son's traditional fairy tales and myths with introductory books on natural history, only to have given the poor boy unprecedented nightmares (see Fig. 5).[107] As the caption explained, this course of action had been embarked upon after reading the preface to Louis Figuier's *World before the Deluge* (1863), which by the time of the cartoon in 1868 had appeared in an English translation that kept its opening 'thesis'.[108] Admitting that some may find his words 'strange', Figuier nevertheless set out his recommendations:

> I assert that the first books placed in the hands of the young, when they have mastered the first steps to knowledge and can read, should be on Natural History; that in place of awakening the faculties of the youthful mind to admiration, by the fables of Aesop or Fontaine, by the fairy tales of 'Puss in Boots,' 'Jack

Fig. 5. 'A Little Christmas Dream', from *Punch* (1868). George Du Maurier's cartoon poked fun at the notion that replacing fairy tale books with natural historical works would be better for childish minds. As his caption claimed, 'we have tried the Experiment on our Eldest, an imaginative Boy of Six. We have cut off his "Cinderella" and his "Puss in Boots", and introduced him to some of the more peaceful Fauna of the Preadamite World ... The poor Boy has not had a decent Night's Rest ever since!'

the Giant Killer', 'Cinderella', 'Beauty and the Beast'—or even 'Aladdin and the Wonderful Lamp', and such purely imaginative productions, it would be better to direct their admiring attention to the simple spectacles of nature—to the structure of a tree, the composition of a flower, the organs of animals, the perfection of the crystalline form in minerals, above all, to the history of the world—our habitation; the arrangement of the stratification, and the story of its birth, as related by the remains of its many revolutions to be gathered from the rocks beneath our feet.[109]

Drawing on his own childhood confusion when learning classical myths, Figuier complained that such tales at best confuse and at worst 'mutilate' children's minds; the more recent addition of fairy tales to the childhood canon had only made things worse. Such posturing could be said to highlight the antagonistic attitudes on the part of leading geological writers such as Figuier towards the entrancing fantastical, and a retreat to dryly technical ground. However, it also betrays an anxiety that the prehistoric parade of monsters was remarkably close to presenting the same kinds of narrative, and characters, and creatures; with correspondingly similar effects. Figuier's claim that the 'marvellous' is 'all things opposed by and contrary to reason' rings hollow when compared with the claims made by, for instance, the *Household Words* authors. Thus these disturbing visions both supported claims for the replacement of fairytales with facts (objects of scientific analysis were now having the same effect as creatures from fairy tales), and also undermined them (these were not a 'better' or 'different' type of story at all).

The palaeontological and mineralogical sciences brought more new creatures to the attention of wider audiences than ever before: as part of the novel science of geology, they opened up vast past vistas—and often vast prehistoric creatures—to public scrutiny and contemplation. In this 'age of monsters', species and specimens would achieve celebrity status, particularly those enshrined on islands in the grounds of the Crystal Palace, part of a scientific and technological culture that seemed capable of ever more impressive results. One way in which their discoverers and analysts harnessed and managed this enthusiasm was through comparison to folklore and fairy tale, positioning themselves as modern dragon-taming heroes, with privileged insights into what was now revealed as the true history of 'once upon a time'. This was no longer a shadowy and mythic temporal span: it had been converted into quantified epochs, named sections identified through creatures and contents, and was the preserve of the men of science. Only the correctly trained imagination, they argued, was capable of such resurrectionary work, able to uncover the true story of life on earth. Comparison to folkloric creatures and fairy-tale plots served to emphasize narratives of progress and modern scientific superiority, and claim that these new scientific fairy tales were capable of everything older stories could do (from providing thrilling adventures and set-piece battles to terrifying young imaginations); yet the very facility with which such comparisons were made betrayed an anxiety about the status of these claims, as part of a nascent science, and as part of an enterprise which dealt with things which could not be seen. Constant recourse to scientific truth, as we shall see in the concluding chapter, is one way in which this anxiety was managed.

If the creatures from folklore and fairy tales were now revealed to have been real-life creatures of scientific study, able to star in superior stories and staged displays, why could this not be extended, others argued, so that by practising the sciences you could be in a fairy tale yourself? In the next chapter, we turn to entomology, to explore how books, images, and children used fanciful characterizations and depictions to investigate the fairy-like creatures at the bottom of the garden: butterflies.

2

Real Fairy Folk

A fairy was sitting up an oak tree in Epping Forest, one summer afternoon in 1885. She had forgotten to put her invisibility cloak on, but her green dress fairly well disguised her within the tree's verdant foliage. The fairy was spying on the passers-by in the glade below, and, striking up a conversation with an owl who had alighted on the bough next to her, wondered at the habits of these strange creatures. Summer in Surrey, the owl informed the fairy, seemed to mean one thing: naturalizing. 'English people', he observed, 'are great worshippers of Nature, and write many guide-books about her.' The fairy was dismayed. 'Does no-one believe in fairies any more?' she exclaimed; and the owl sadly detailed how the belief had indeed been 'snuffed out by the scientific men'. As luck would have it, a specimen of the 'scientific men' themselves just then entered the glade, tall and thin, wearing spectacles and carrying a net and knapsack. Having learnt about their existence only moments before, the fairy's interest was piqued, and she leant forward to see more of the habits of this novel species. She watched, intrigued, as the scientific man took some small bottles and a tin box out of his bag, and began to skim the surface of the pond with a ladle, and collect its contents. He peered into his phials, while the fairy wondered what on earth he was up to. Shouldn't he be fishing instead? Then, all of a sudden, he seemed to be extraordinarily pleased. He let out a cry—'Stephanoceros!' Unfortunately this startled the hidden fairy, who fell from the bough down towards the water. Attempting to fly away from the stagnant pond, her futile fluttering drew the attention of the scientific man, who abandoned his previous specimen to bottle, instead, this extraordinary creature. Never, he thought, had he seen 'such a remarkable butterfly'.[110]

A conflation of the fairy and the butterfly, as seen here in Andrew Lang and May Kendall's story *That Very Mab* (1885), was commonplace in Victorian Britain. Fairies made frequent appearances in Victorian entomology, as fanciful works played on the supposed similarities between insects and fairies, from size to wings to movement to ephemerality.[111] Indeed, fairy tale scholars see the nineteenth century as the crucial time period when these similarities were enshrined; when fairies shrank to Tinkerbell proportions, became be-winged, became benign.[112] Whether in high art or satirical commentary, conventions of giving fairies butterfly-like wings, cricket-like voices, and a mayfly existence connected the fabulous creatures to inhabitants of the insect realm, not just as victims of mistaken identity. But beyond such aesthetic similarities, tales and images of fairies can also be connected to hands-on entomological practice, which reached new audiences in the period. Searching and collecting, pinning, bottling, displaying, observing, analysing, and discussing: entomology was an increasingly popular participatory science in the nineteenth century. It could be identified with a fabulous story itself. It could blur the boundaries between people and fairies and insects. And, as in the case of the Surrey naturalist, going on entomological adventures could be the best way to encounter the real fairy folk.

A story without an end

Episodes of Insect Life by 'Acheta Domestica' was an introductory entomological work published in three volumes from 1849 to 1851.[113] Hiding behind the narrative persona of a cricket ('Acheta Domestica' is the feminized scientific name for the species), in this series of books author Louise M. Budgen constructed an insect chronicle for her readers. The chapters in the book were organized around the seasons of the year, but jumped between different narrative forms: poem, parable, fairy tale, anecdote, exposition.[114] The chapters made the most of the high-quality production values of what was a relatively expensive gift-book, as they opened with a realistic illustration of the entomological subject-matter of the chapter, and closed with a whimsical vignette that converted insects into human-sized characters.[115] The vignettes placed these insect characters alongside the latest in Victorian life, furnishing them with contemporary objects from guitars (to illustrate insect senses and communication) to balloons (mimicking insect aeronautics), to acorn phaetons. In the third volume of Budgen's book, one illustration depicted 'The 'Exhibition of the Industry of All Insects' (see Fig. 6), referencing the Great Exhibition of the Works and Industry of All Nations that was taking place the very same year.[116] For best-selling natural history author the Reverend J. G. Wood, who edited a later edition of *Episodes*, the line-drawn insect characters that disported themselves in the margins of the book were themselves as worthy of microscopic analysis as the insects whose study Budgen had aimed to inspire:

> Most of the drawings must be examined, as the insect itself must be viewed, with the aid of a magnifying glass; and not until this is done, will the singular truthfulness of their execution be seen.[117]

Here Wood quite deliberately blurred the lines between actual, fantastical, and represented objects: reading an entomological work was hence a means of trying out practices of observation. Crucially, the images were held up as being of 'singular truthfulness': based on accurate drawings of real

Fig. 6. 'We Challenge all Nations!', 'Acheta Domestica', *Episodes of Insect Life*, Vol. III (1851). The inventive illustrations to *Episodes of Insect Life* depicted its entomological subject-matter as inhabitants of contemporary Britain. Here, wasps and ants are shown picketing 'The Grand Exhibition of the Industry of All Nation', a play on the 1851 'Great Exhibition of the Works and Industry of All Nations': individual creatures hold signs that emphasize the 'Industry of All Insects'.

specimens, they were hence an educational as well as often comic guide to their audiences' subsequent glimpses at insect life.

In her preface to the successive yearly volumes of the book, Budgen carried on a conversation with her implied reader, elaborating on the reasons why she had written her tales in an idiosyncratic way. Budgen did not just want to teach insect facts to her readers: she had, she explained, an 'ulterior and more useful design', that 'of cultivating the rudimental seeds of systematic investigation'.[118] She wanted to enlist her readers as part of the new audiences for all kinds of natural historical study, which were growing rapidly at mid-century, and which had witnessed successive waves of fashionable pursuits: shells, ferns, and, soon, seaweed.[119] The popularity of these pursuits was one reason why Wood could assume his readers had a microscope with which they could examine the book's illustrations. Butterflies and dragonflies were manifestly attractive creatures to study, and had long been deemed suitable scholarly companions for young women; but Budgen also wished to make the uglier, fiercer, grubbier inhabitants of the insect world 'objects of liking'.[120] This was, she argued, 'the best preparatory step towards making them subjects of learning'.[121] Fanciful illustration and fictionalized narratives were two important literary tools she deployed in pursuit of this goal, as she demonstrated the crucial importance of the particular ways in which potentially unappealing scientific subjects were dressed up for new audiences. (Quite literally, in the case of many of her illustrations: one image, for instance, showed three 'painted ladies': butterflies taking tea in the garden—see Fig. 7.) By writing in miscellaneous literary styles, Budgen self-consciously drew attention to the choices that had to be made by introductory authors who sought to recruit new participants to their discipline: that there were many types of scientific stories, including fairy tales.

Budgen also revealed her self-awareness about the important role of story-telling in the sciences when, at another point in the preface, she compared the pursuit of entomological science—in fact, of the entire field of natural history—to casting oneself in a fictional narrative. Whereas both author and reader, she argued, experienced 'a feeling of regret at having *done with*' the 'persons and scenes' of literary fiction or history, 'it is never thus with the objects drawn from the world of nature'. As she went on:

> With these neither writer nor reader are ever called upon to part.... to both, when once endeared by awakened interest, they are ever present... Completed works on natural subjects? There are no such things! The most scientific of them is but the commencements of 'a story without an end;'—the least so (this among them) is but an invocation to begin and read it![122]

Here, Budgen made an explicit comparison between story-telling—indeed, living in a tale—and hands-on scientific

"Sipping their cups of dew."

Fig. 7. 'Sipping Their Cups of Dew', 'Acheta Domestica', *Episodes of Insect Life*, Vol. II (1850). Butterflies are shown taking afternoon tea, in another illustration from *Episodes of Insect Life* that treated its insects as humans participating in contemporary activities; here, folded wings become the skirts of dresses, antennae rather alarming headgear.

practice. Her literary analogy played on the religious analogy underpinning much contemporary natural theology that was often exploited by scientific writers, for instance in the elementary geological books we met in the first chapter, and in the evolutionary works we will explore in Chapter 4. This cast the surrounding environment as 'The Book of Nature', God's 'second book', which, like the Bible, his first book, could be read closely for moral and intellectual enlightenment. Metaphors of reading 'sermons in stones', commonplace in geological writings, were reworked in entomological texts, which, rather, tended to stress providence and design. Margaret Gatty's *Parables from Nature* (1855), an overtly religious work, claimed inspiration from Thomas Browne's *Religio Medici* (1642), who both emphasized the 'two books' analogy, and also credited the insect world with having added a spiritual dimension to his wonderings: the 'strange and mystical transmigrations that I have observed in silk-worms turned my philosophy into divinity'.[123] Gatty similarly used insect metamorphosis in the first of her *Parables*, as an allegory for eternal life. She reasoned that, were a caterpillar 'capable of contemplating its own existence, it would reject as an impossible fiction' the fact that it would transform into a butterfly.[124] Similarly, human beings were cautioned against rejecting the possibility of an analogous angelic ascension because of our earthbound existence; and, like the caterpillar in Gatty's tale, were given 'A Lesson in Faith'. This is, indeed, in striking contrast to Hans Christian Andersen's fairy tale 'The Butterfly', which had a rather different moral.[125] Chronicling the search of a butterfly for an appropriate floral bride, it taught suitors not to be too picky, lest they miss out on a spouse all together.

Worse, the butterfly ended up, like Queen Mab in our opening story, being caught by an enterprising entomologist. Stuck in a cabinet, for the newly pinned butterfly practising entomology was here the end, rather than the beginning, of his story.

Budgen's *Episodes* departed strikingly from Gatty's and Andersen's stories in its far more in-depth presentation of the insect world. She did not name merely a generic 'butterfly' and 'caterpillar', which were used as metamorphosing ciphers for more general moral lessons: just as her insect persona had a precise expert moniker, she also tailored her tales to specific attributes of insect species. For instance, her story 'Butterflies in General' played upon the analogy between butterflies and flowers, before moving into a more active pursuit of a particular butterfly, who was given a character appropriate to his common name:

> Let us follow one to the garden. Behold him seated on his velvet cushion, the corolla of an aster or a single dahlia...His long spiral tongue has hitherto lain coiled betwixt two side appendages, but now unrolling, he plunges it to the bottom of a chosen chalice, then partially recurves, and indraws his honied draught through the tubelike sucker. Again and again, he quaffs like an 'Alderman'† as he is. We know him by his bulk and the richness of his furred and velvet robes, scarlet and black, relieved with white. But see how the rights and pleasures even of an Alderman Butterfly are open to invasion! Look at that impertinent prying 'Argus',‡ tired of his rustic fare in heath or meadow...Down he lights and seats himself beside the dahlia table, an unbidden guest...But, can it be possible? the little Argus, not content with a dinner upon sufferance, has

actually become the assailant of his unwilling host. He closes
his blue wings... and then tries with his pigmy body to dis-
lodge, by shoving, the corporation of the Alderman...
   † *Vanessa Atalanta*, Alderman or Red Admiral Butterfly
   ‡ *Polyommatus Argus, P. Alexis*, Common Blue Butterfly...[126]

The specific coloration, naming, and behaviours of the two
different types of butterfly were contrasted by their roles in
this mini-drama, which Budgen spins out into a little story.
This miniature tale was just one example of the many
endless stories to be found in nature, which, as she claimed
elsewhere in the book, are there to be noticed:

> while the Thousand and One Nights of the far-famed Schere-
> zade [sic] are in everybody's memory, the 365 days of the year,
> each with its tale within tale of wonders ever new, go round
> unheeded or unheard. (p. 145)

With her chronological presentation mimicking the quotidian rhythm of the *Arabian Nights*, Budgen, perhaps, intended
to become Nature's Scheherazade.

The adventures of Madalene and Louisa

These comparisons between entomological practice and
story-telling did not just take place in the fanciful mind of
a pseudonymous house cricket: actual Victorians, young
and old, made such connections themselves. For instance,
between the ages of 12 and 16, the upper-middle-class

Victorian teens Madalene and Louisa Pasley chronicled their real and imagined insect-hunting adventures in a playful illustrated album. As they put it, they 'preferred chasing beetles and butterflies to lessons in the schoolroom'.[127] This beautiful picture-book presented the sisters as middle-aged spinsters, encountering many different types of insect that lived near their home. Oil beetles, glow worms, privet hawk caterpillars, an 'unknown larva', and even 'sportive ichneumon flies' jostled for attention on the book's pages, alongside the sisters and their pets, the surrounding English scenery, and everyday pastimes from boating to a ball. Overcoming the disapproval of those who surrounded them, Madalene and Louisa were cast as heroines, transcending the tribulations of bursting beetles, recalcitrant caterpillars—and often even more troublesome humans—to achieve entomological success. The Pasley sisters lived and worked, wrote and drew, in the mid-nineteenth century, and their surviving work reveals how readily hands-on entomological pursuits could be thought of as a story, and put into fictionalized form.

Madalene and Louisa were not the only characters in the story of their life: they played with scale to show insects as worthy antagonists, inflating them to human-size. In the series of imaginative scenarios sketched out in the album, many activities that might have formed part of regular entomological practice were also played with for comic effect. The perfectly possible encounter with a 'rare larva', when enhanced with illustrations, became a fantastical mini-narrative as much as an experimental mishap:

> We discover a beetle larva in a dead tree.
> With great effort it is secured
> in a collecting case.
> It however bursts the ends out
> and another method has to be adopted.[128]

Insects were also lassoed with fishing wire, bitten by dogs, forcibly pushed through windows, or even converted into carriagemen, who 'sprang into the Phaeton and drove off without us'. Such active devotion to the natural historical cause was all-consuming: the sisters would often sneak out early, stay up all night watching a chrysalis hatch, or 'make DARING NIGHT EXPEDITIONS'. This, understandably, had an adverse effect on some of their other pursuits, often to the annoyance of those trying to teach them:

> We explained to a series of daily governesses that we would rather study ENTOMOLOGY than ARITHMETIC—but none of them was interested in beetles and all of them persisted in setting us SUMS.

Entomology was presented as a far more enticing activity than mere mathematics: something that encouraged an active engagement with the surrounding world, and taught the kind of precise factual information that only enhanced daily life: in 'the light of the lantern, quite ordinary moths and beetles looked LARGER and more interesting'.

Such an affirmation of the superiority of those who entomologized was compounded with the introduction of Madalene and Louisa's elder sister, Georgie. A constant irritant to her younger siblings, she had been sent away to school, where she had become boring and 'bitter'. Worst of all, instead of admiring insects she fed Madalene and Louisa's beetle specimens to her pet goldfinch, 'Mr. Rowley'. She preferred to moon over their drawing instructor, who rowed them out on a lake to see the sunset: 'He wore his top hat and recited a poem which Georgie thought ROMANTIC.' When forced to go out bug-hunting, she always put a dampener on their enthusiasm, which, again in an overt use of a literary trope, was reflected in a meteorological pathetic fallacy: 'it was bound to RAIN.' However, perhaps Georgie had been too hasty in rejecting the romance of natural history: at the time there was also matrimonial potential to be found in the insect world. For instance, in 1845 Samuel Stevens, a member of the Entomological Society (founded in 1833), claimed to be 'on the lookout for an entomological wife', as she would be 'useful as well as ornamental'.[129] It was common for collaborative couples to work together on scientific pursuits throughout the nineteenth century; many spouses contributed often undercredited expertise in collecting, identifying, drawing, or writing about natural historical specimens.

Many of the album's more spectacular interactions between text and image—which to modern readers resemble graphic novels—were drawn from conventions of the contemporary comic press, as well as from other popular leisure pursuits for upper-middle-class women and children like the Pasleys, such as botanical art, and scrapbooking.[130] From the 1820s there had been a surge of interest in scrapbooking and, by the third quarter of the century, other young Victorians would combine insect illustrations with the new photographic *cartes de visite* in their scrapbooks. Often these were subversive reworkings of familiar members of the household or famous figures of the day.[131] For Madalene and Louisa, the album served a similar function; as when they presented a thought-experiment of what would have happened if the disapproving drawing master, Mr Mitchell, 'were to join in' on their adventures. She 'drew what might happen to him if he did':

1. Mr. Mitchell begins to sketch on (as he thinks) a log of wood.
2. It proves to be a 'looper' caterpillar and towards evening it rises erect to sleep.
3. L. & M.P. procure a ladder.
4. The caterpillar feeling tickled doubles half up.
5. Further endeavours at rescue.
6. By a sudden jerk the catterpillar [sic] throws Mr. M. to the ground.
7. He is carried home by his pupils on the ladder.

This story had an unfortunate ending, when Mr Mitchell accidentally saw the drawings and 'his WHISKERS went stiff with rage'; the sisters, he claimed, 'would never learn to draw...anything serious'.[132] Yet it also helped the narrative on to its conclusion, when their father finally despaired of finding accommodating instructors, and let his two younger daughters 'entomologise as much as [they] liked with no one to bother [them]', catching 'a great number of rarities' for their collection. There was a happily-ever-after of a sort, too: Madalene wrote more traditional works on butterflies, including an 1862 manuscript volume.[133] However, as 'The Adventures of Madalene and Louisa' concluded, 'nobody liked them as much as THE ALBUM.'

Fairy Know-a-Bit

So far, fairies have had rather an oblique existence in this chapter. But in the two books I will discuss now, the eponymous siblings *Fairy Know-a-Bit* and *Fairy Frisket* moved into a starring role. Both were works from the pen of Charlotte Maria Tucker, who, as A.L.O.E. ('A Lady Of England'), wrote very many books for children, on needles or rats to India.[134] In *Fairy Know-a-Bit; or, a Nutshell of Knowledge* (1868), she used an insect fairy to teach children about their surrounding world. In its sequel, she introduced other types of insects themselves, through bodily transformation and magical devices, and another fairy guide (see Fig. 8). As the sequel's preface stated, readers would 'scarcely expect' to encounter Know-a-Bit or Frisket on walks around their

Fig. 8. 'The Two Fairies', from A.L.O.E., *Fairy Frisket; Or, Peeps at Insect Life* (1874). An illustration of the meeting of 'The Two Fairies' revealed the differences between the eponymous stars of Charlotte Tucker's two books: whereas Know-a-Bit has taken on human clothing and education, crushing his wings under an academic gown, Frisket's insect-like characteristics, including her butterfly wings, are still in evidence.

gardens, springing 'from under a fern-leaf', or sitting 'on a hawthorn spray'; but they 'very probably' would encounter their insect brethren.[135] Tucker's choice of teacherly fairies may well have been inspired by characters in a story told by a character in Catherine Sinclair's 1839 *Holiday House*, in which 'two magnificent fairies' named fairy 'Do-Nothing' and fairy 'Teach-All' visited an idle boy, Master No-Book.[136] Charles Kingsley's *Madam How and Lady Why*

(1870) was also a likely source of inspiration with its fairy-like instructresses, versed in the secrets of the natural world.[137]

Fairy Know-a-Bit was to be found lurking in the library at the subtly named 'Fairydell Hall' (see Plate 2). The tale starred two 'human specimens': rich, idle, and selfish Philibert Philimore, aged 7, and the virtuous, poor, yet knowledgeable Sidney Pierce. Philibert's first meeting with Know-a-Bit, and the fairy's first description, emphasized his insect-like attributes; but also how he had attempted to disguise himself as a human:

> Seated on the volume, quite at his ease, appeared a tiny figure, not six inches high, dressed like a student, in cap and gown, with wee dots of spectacles on his nose, and a grand beard, nearly an inch in length, which reached his little girdle! The figure had as a pen behind his ear a quill from a humming-bird's wing, and at his girdle hung an ink-bottle about the size of an elder-berry. His eyes, not quite so large as those of a robin, but a good deal brighter and merrier, twinkled through the tiny spectacles, which looked like diamond dew-drops set in a single thread of gold! An elegant little creature was this to behold, as he sat there with a tiny white wand in his hand, hung with wee silver bells, which tinkled when he moved it. Philibert was astonished and delighted, but yet a little frightened, for he had never before in his life seen anything so pretty or so strange![138]

In *Episodes of Insect Life*, Budgen had assumed an entomological persona: that of a cricket (also depicted next to the library shelves); but in this story an insect-sized fairy was dressed up as a person, more like the closing vignettes to

Budgen's chapters. The representation of insects as people and as fairies drew on the tradition, from Aesop onwards, of using the creatures as the basis of moral tales, and also the popular rhyming books written to mimic the famous early-century *Butterfly's Ball*, which have been called 'papillonades'.[139]

Know-a-Bit went on to introduce himself to the startled boy, telling how he was 'once a fairy', wearing 'the petals of flowers, or spoils from the butterfly's wing', but now:

> times have changed—and so have I. A railway now runs right through the valley which was our favourite haunt—there are engine-lights instead of the glow-worm's...! Education is now all the fashion, and fairies, like bigger people, are sent to learn lessons at school.

Echoing the common narrative that I disputed in the Introduction, and that had been emphasized in *That Very Mab*, Know-a-Bit blamed steam-driven industrial modernity for chasing away the fairies from nature. Know-a-Bit had 'taken to books' with the invention of printing, and transferred to a new bibliographic home. In *Know-a-Bit's* sequel, his sister, Frisket, took him to task for hiding out in such a 'den':

> 'Ay, in a *room!*' exclaimed Frisket with scorn. 'You choose spectacles instead of free wings; books instead of leaves and mosses and ferns and flowers! You like to hear the mouse squeaking behind the old wainscot, instead of the lark singing in the air!'[140]

'[M]imicking man', in Frisket's words, Know-a-Bit had even 'crushed down' his wings 'under that black gown'.

The works emphasized the important role both books ('I will rather leave my young readers to find out for themselves all that the fairy might show them. Knowledge, like Know-a-bit, lives in books') and fiction ('is not impossible for mind to go where body cannot enter, when fancy is powerful') had in educating childish audiences. The works made the most of 'fancy' with their attribution of magic powers to the two insect characters, powers deployed to enhance their role as educational characters. Know-a-Bit could use the tassel on his cap to 'appear and to speak' to mortals; and with his wand and a magic mirror could conjure instructive images and reveal the hidden properties and histories of objects at the house. By the aid of Frisket's magical pomatum ('the newest invention in fairy-land'), the boys' minds were transferred into insect carapaces and could go to a butterfly's ball (their human forms safely sleeping back in the bedroom); they could 'buzz through a hive as bees, or roam through underground passages as ants, or bury themselves like beetles, or fly through the air as gnats'. Frisket used her insect-like properties—the small size of her body, and its quick, darting movement, her 'dragonfly' wings—and her magical powers—the pomatum charm—to aid in the introduction of the worlds and types of insects in the garden and woodlands that surrounded Fairydell Hall.

Real Fairy folks

Everybody knows you can find fairies at the bottom of the garden, as well as up oak trees in Epping Forest. Darting and whirling, these magical winged creatures are ephemeral and elusive, glimpsed out of the corner of your eye, or heard in whispers on the breeze. This chapter has explored how many Victorians sought to enquire whether insects were, in fact, the real fairy folk. Just as fairies were given insect attributes, so were insects converted into fantastical creatures, starring in their own natural historical tales, guiding introductory education in the sciences, and recruiting new participants in the unending story of entomology, which promised a happily ever after. By the late nineteenth century, then, the be-winged fairy was a familiar icon; and also, as we have seen, often referenced in introductory scientific works and practice, particularly the common invocation of natural historical fairies (notably, insects and flowers). But in 1887 all of these tropes were brought together in one elementary chemical text for American children by medic, author, educator, and Methodist worker Lucy Rider Meyer (see Fig. 9). She invited her readers to join a fictional family in their 'explorations in the world of atoms', under the guidance of an avuncular professor, who taught of the fairyland of chemistry that could be entered through elementary investigations.[141]

SOME OF THE REAL FAIRY FOLKS. *Frontis*

Fig. 9. 'Some of the Real Fairy Folks', frontispiece to Lucy Rider Meyer, *Real Fairy Folks, Or, The Fairyland of Chemistry: Explorations in the World of Atoms* (1887). The introductory image to Meyer's work reveals the winged fairies interacting with scientific equipment, in a clear statement of how her book would explain chemical processes.

Like many of the texts aimed at a young audience, *Real Fairy Folks; Or, The Fairyland of Chemistry: Explorations in the World of Atoms* (to give the work its full title) was structured around an appealing domestic narrative, with the author's expertise couched as a knowledgeable adult character, and the readers of the work elided with the children, Jessie and Joey, who were learning all about the fascinating new subject. Its central conceit was that atoms should be thought of as fairies, whose behaviour, dress—and even limbs— reflected their chemical properties. For instance, different groups of elements were sorted into 'firms of fairies' or 'cousins amongst the fairies', with similar properties.[142] The different states of matter were introduced as reflections of how active the fairies were. There was an emphasis on the 'work' fairies do in the world, for instance the use of chlorine in disinfecting hospital wards (where 'a lot of Chlorine fairies' were 'let loose') or in bleaching linens (where 'a few men, aided by the tiny hands of the fairies, do in a single day the work that used to require thousands of workers and months of time').[143] The superior nature of these chemical fairies to the common or garden varieties was clearly demonstrated:

> 'I didn't suppose fairies ever did any real work,' said Jessie, with a little disappointment in her tone. She was thinking of the fairy stories she had read.
> 'Story-book fairies do not, but these *real* fairy-folks are not afraid of hard work.'[144]

The text brought together different ways of representing chemical structures and bonding—a crucial concern of the nineteenth century, which had only recently witnessed the invention of the first molecular models, now specially produced but at first based on table croquet sets.[145] Central to these concerns was what kind of representation this would be: for example, would the atomic arrangements reflect their real-world positions or not? Alongside the conceit of the fairies, revealed most vividly in the illustrations to the book, which also depicted basic chemical equipment, attendees at a chemical lecture, and occasionally the children and their uncle, other means of learning chemical properties were also incorporated:

> their uncle proceeded to test them by asking them how they would tell the story of the union of chlorine and hydrogen to make hydro-chloric acid. With a little prompting, they said, and the Professor wrote it on the board:
>
> **H + Cl = HCl.**
>
> 'But, how's this? We've made the Hydrogen fairy stand alone, and the Chlorine too, while the fact is, they are never a moment alone—must have something to cling to, always. Let us try this way:'
>
> $H_2 + Cl_2 = HCl.$
>
> 'You can't do that,' said Jessie decidedly, 'because one of each kind gets lost.'
> 'Sure enough. Well, how's this?'
>
> $H_2 + Cl_2 = 2HCl.$
>
> There was no fault to be found with this...[146]

The notion of the fairy atoms was combined with the more usual use of chemical symbols, to balance the sides of the equation so that fairies were not 'lost', but also retained the fact that these particular fairies were 'never a moment alone'. Importantly, the Professor had talked about these equations as if they were themselves a narrative, running left to right across the page ('tell the story of the union'). Just like the standard plots of the fairy tale ('Frog + Kiss = Prince'; 'Orphan + Fairy Godmother + Pumpkin + Glass Slipper = Happily Ever After'), this 'union' was one that could be talked about in storified terminology, introducing a temporal as well as a universal element (see Fig. 10). Just like Budgen, who talked about scientific practice as an 'unending' story, or Madalene and Louisa, who cast themselves as characters in scientific stories, it was to literature that the sciences were compared: even the facts of scientific processes like this.

As Wood had commented in Budgen's *Episodes*, it was crucial to stress that these overtly fanciful stories were in fact especially factual and precise works, which were supposed to convey accurate information and represent plausible domestic experiments. In a preface to parents, Meyer declared that the book was 'true to chemical fact and principle', and that it was 'an honest effort' to make children 'love the beautiful science of Chemistry'. Meyer had previously taught chemistry at McKendree College, Illinois, a Methodist college that had been established in 1828, and

HYDRO-CHLORIC ACID.

Fig. 10. 'Hydrochloric acid', from Lucy Rider Meyer, *Real Fairy Folks, Or, The Fairyland of Chemistry: Explorations in the World of Atoms* (1887). Images were interspersed throughout Meyer's text, showing elementary experiments and representations of chemical processes, as well as more playful depictions of objects from the story. This illustration of hydrochloric acid showed the substance as a be-winged fairy, much like contemporary representations of insect or flower fairies as humanoid creatures with wings.

used her experience to advise how best to achieve this aim. She suggested that adults read aloud the book with children, 'not too much at a time, winter evenings and summer vacations': as with the goings-on at Fairydell Hall, this was to be an education that took place outside of the school-room. Crucially, she urged the need for 'imitative experiments' that would mimic those in the book, blurring distinctions between actual and represented practice.[147] Many of the demonstrations in the book did use everyday objects—candles, sugar, vinegar, matches, mince pies—facilitating this process.

For Meyer, then, fairies were not to be found through interactions with the insect world in the hives and hills at the bottom of the garden; nor even through imagined conversations with insect guides in the library. They were, rather, the constituent parts of everything in the surrounding world; the fundamental building blocks of matter were best conceived of as marvellous miniature creatures, invisibly going about their work. Even a humble molecule of water was best represented as a 'fairy picture' in which an atom of oxygen held hands with two hydrogen atoms (see Fig. 11).[148] Drawing on contemporary representations of fairy-kind and angelic beings, with flowing hair and liquid gowns, radiant headdresses and benign expressions, the common substance was rendered unfamiliar, and yet comfortingly recognizable as something that was both of and beyond nature: exactly the spiritualized message Meyer

FAIRY PICTURE OF WATER.

Fig. 11. 'Fairy Picture of Water', from Lucy Rider Meyer, *Real Fairy Folks, Or, The Fairyland of Chemistry: Explorations in the World of Atoms* (1887). A water molecule in this image becomes a trio of fairies: the oxygen fairy in the centre holds hands (Meyer's metaphor for making chemical bonds) with two hydrogen fairies.

sought to communicate to her readers. The next chapter will go further into the peculiar thrall in which drops of water held many in the nineteenth century, as they were revealed as fairy suitors, miniature worlds, or even monster soup.

3

Familiar Fairylands

'THE THINGS I HAVE SEEN IN TAPIOCA PUDDING...!'

G. H. Lewes (1858)[149]

The opening pages of Agnes Catlow's 1851 introductory work on microscopy gave a curious set of directions.[150] Her readers were instructed to 'fancy themselves spirits, capable of living in a medium different from our atmosphere': only then could they accompany the author on a journey through the 'wonderful brazen tunnel, with crystal doors' at one 'entrance', and a matching set of smaller 'portals' at the other.[151] Catlow's narrator took time to praise the 'time and labour' that had brought the 'peculiar' gates to 'perfection', and told of how 'a spirit, Science', was their custodian.[152] Revealed behind these opened doors was an avowedly 'new world', certain to 'astonish' and 'bewilder' those who had never travelled there before.[153] Luckily, the narrator was on hand to reassure and guide the reader through the subsequent encounters with novel creatures both inanimate and animate, with bizarre compound, ribbon-like, or globular forms.[154] Theirs was only to be a 'transient' visit, however, as the readers were soon invited to return to their 'own world', restricted to 'beings like [them]selves'.[155] Though this opening may suggest a fantastic voyage, Catlow's readers had not in fact fallen down a rabbit-hole into wonderland: they had instead been looking down a microscope, at a drop of water.

Drops of water, as in Catlow's eponymous book, formed one of the most popular subjects with which readers were baptized into the scientific community in Victorian Britain. An almost everyday occurrence to many children growing up in these drizzly isles, their closer examination provided a simple and clear example of the invisible wonders of the surrounding world. Importantly, these were wonders that could be revealed by the power of scientific instruments and investigations. For everyone, from famous fairy tale tellers including Hans Christian Andersen to introductory science-writers such as Agnes Catlow and Arabella Buckley, as well as scare-mongering journalists and satirical cartoonists, the lessons of the water droplet embodied the best introduction to scientific objects and practices. Fairy stories written about drops of water in the mid-nineteenth century show how everyday substances could be revealed as homes to magical creatures, miraculous forces, and dangerous monsters.

These stories connected their readers to critical contemporary debates over water purity and public health, to newly affordable parlour microscopes, to educational experiments using objects gleaned from house and garden, and even to a potential love-affair with a water-sprite. In their pages the familiar domestic environment was recast as a fairyland of science waiting to be explored by its inhabitants, if they only opened their eyes, and their minds. The powers of science, such stories showed, were all around, from the incessant boiling of the tea-kettle on the stove, to the secrets hidden inside George Henry Lewes's tapioca pudding.

The world in a drop of water

In 1848, many European cities were in turmoil: as the 'year of revolutions' ran its course, from Paris to Scandinavia to Vienna, teeming and often violent groups agitated against the ruling classes in waves of successive revolts.[156] A similarly vicious world could also be found that year inside a drop of water, as Hans Christian Andersen's fairy-tale 'A Drop of Water' detailed how its aged protagonist, Kribble-Krabble, tricked a visiting magician into believing that he somehow saw 'Paris, or some other great city, for they're all alike' when looking through a glass.[157] In the 'drop of puddle water' placed under magnified view, the visitor had witnessed a scene unfold before him that greatly resembled the internecine struggles on Europe's streets that were being reported daily in the pages of the newspapers:

> It looked really like a great town reflected there, in which all the people were running about without clothes. It was terrible! But it was still more terrible to see how one beat and pushed the other, and bit and hacked, and tugged and mauled him. Those at the top were being pulled down, and those at the bottom were struggling upwards.[158]

Andersen's depiction of the once 'great town' converted into a 'terrible' spectacle, emphasized by the repetition of the word, revealed anxieties about social reform and the potential upsetting of the established order. Just as in many of the civic demonstrations, the upper classes, aristocracy and monarchy ('Those at the top'), were threatened ('pulled down'), as the lower classes sought greater rights, influence, and representation ('struggling upwards', in a phrase reminiscent of Lamarckian transmutation). The watery nature of this magic mirror was punned upon as the great town was explicitly 'reflected' in the drop. The reason for the magician's name also became clear, as the onomatopoeic phrase 'a kribbling and a krabbling' emphasized the swarming and squabbling of the creatures themselves.

The tool that had given Kribble-Krabble, and his deceived visitor, the ability to peer into this miniaturized version of their own society was a magnifying glass. Simple lenses for magnification had been available for centuries: associated with natural magic, the story's themes of trickery and deception fit with older conjurings of grotesquely enlarged, deformed, or recoloured forms. Yet the story also referenced and used modern techniques of microscopic slide preparation—staining dyes such as carmine that the narrator renamed 'witches' blood'.[159] By the 1840s, home microscopy was an achievable pursuit for the middle classes, an appropriately rational, devout, and curious activity. The price of microscopes had dropped substantially throughout the 1830s, and some of the worst defects of earlier instruments, in particular the refractive distortion known as chromatic aberration that lent an otherworldly tint to the magnified objects, were being remedied.[160] Kribble-Krabble was not alone in inviting visitors to look down microscopes. Many books for home pursuits were produced around mid-century, including Philip Henry Gosse's *Evenings with the Microscope* (1859) and J. G. Wood's bestselling *Common Objects of the Microscope* (1861); moreover, microscopy was central to a range of scientific interests, from mineralogy to entomology, botany to medicine.[161] Pieces of rock, petals of flowers, and even parts of the body were transformed into wonderful sights as they revealed their secrets through the lens. Microscopes formed a key part of scientific entertainments outside the home, too: as can be seen in contemporary engravings of metropolitan events, which portray serried ranks of microscopes at evening conversaziones, with circular images thrown up on the wall, mimicking microscopic sights, and with large audiences using these instruments to glimpse other realms.

The keen audiences for microscopic soirées formed just one manifestation of a much wider culture of spectacular shows, particularly in fashionable London.[162] At several West End venues, audiences could see a view of the capital from the spire of St Paul's cathedral, vistas of the ancient world, biblical stories brought to life, or exhibited peoples from across the globe. The latest developments from the Continent and burgeoning Empire were depicted on the big screen with magic lantern shows, much like the early cinema newsreels that preceded black-and-white films. By this point the magic lantern, a simple projecting device, was an established technology: some more elaborate incarnations could even switch between slides to give cinematic effects, such as the impressive dissolving view; or had gears or movable parts, which translated into moving images. As well as militaristic, religious, and geographical images, the sciences were a popular topic. These included the giant projections of the oxy-hydrogen microscope, in which real water was put into slides in specialized 'cages'.[163] Later in the century, books designed for domestic audiences included schematics for these 'cages'.[164] When enlightened, magnified, and thrown across the room onto a large screen, the previously transparent substance was revealed to be full of life, home to bizarre and wriggling creatures (see Fig. 12). Just as with the battle-vistas or exotic lands audiences would

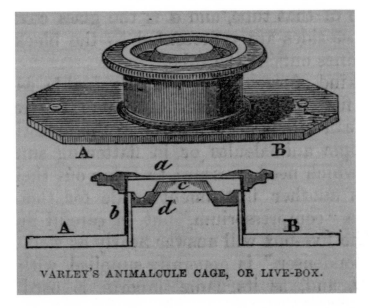

VARLEY'S ANIMALCULE CAGE, OR LIVE-BOX.

Fig. 12. 'Varley's Animalcule Cage, or Live-Box', from J. G. Wood, *Common Objects of the Microscope* (1861). The schematic demonstrates how water, and its minuscule inhabitants, could be incorporated into the magic lantern apparatus. The inclusion of such illustrations in best-selling works such as this, intended for the home market, reveals a widespread audience for microscopical instruction: by the 1860s such investigations had moved from public demonstration into middle-class homes.

have encountered at other shows, these violent scenes showed warring beings and strange worlds, made visible for the first time. Perhaps, like Andersen, they would have brought together these two types of news from another realm: distanced (as the projections were from the audiences) from everyday reality, but very much affecting it.[165]

Those bizarre wriggling creatures that the magic lantern magnified to 'terrible' effect were known as 'animalcules', as they had been dubbed in the seventeenth century by pioneering Dutch microscopist Antonie van Leeuwenhoek. They also appeared down the microscopes at fashionable evening soirées given by organizations such as the Geological Society, as well as at private gatherings. The Sussex surgeon and palaeontologist Gideon Mantell, for instance, was introduced to the amazing array of minute organisms one could discern down the microscope at a dinner party in 1844; and in March 1845 took some animalcules, stained with carmine, to a Geological Society soirée.[166] He was so entranced by this practice—and wished to bring it to greater attention—that he quickly wrote a successor to his popular geological primer *Thoughts on a Pebble* (1836), which appeared as *Thoughts on Animalcules; or, a glimpse at the invisible world revealed by the microscope* in 1846.[167] In the book's opening pages, as had been evident in his subtitle, Mantell was keen to compare his subject-matter to previous traditions of hidden water-creatures. He began:

> In every country and in every age, a belief in the existence of beings invisible to the mortal eye has more or less generally prevailed; and the air, the earth, and the waters have been peopled by ideal forms, invested with natures and attributes partaking of the characters of the minds from which they emanated. Hence sprang the Gnome of the mine and the cavern; the Goule [sic] of the charnel-house and the tomb; the beautiful mythology of Fairy-land...[168]

Mantell included an image of the animalcule he dubbed the 'water-nymph', reifying this connection between the creatures of myth and the creatures of microscopic perspective. Like others in the nineteenth century, he made a clear connection between fairy tale inhabitants and narratives, and the new discoveries of the sciences. For Mantell, these animalcules could clearly be ghoulish or 'beautiful', harmless or threatening, depending on one's perspectives. For another commentator, however, the swarming masses found in the 'muddier portion of the water' were 'Vathek and his companions in the Hall of Eblis'; a reference to a popular contemporary oriental tale, inspired by new versions of the *Arabian Nights*. Like the characters in the story, the 'moving creatures' were 'in a similar manner avoiding each other, as if repelled by some innate force'.[169]

In a poem prefaced to Catlow's *Drops of Water*, preceding the imaginative journey down the microscope with which this chapter began, the beauty of the waterdrop world rather than its horror was highlighted, as she lauded the tiny animalcules glimpsed within the droplet, those 'creatures beautiful and bright, | Disporting 'midst its liquid light'.[170] In order to help her reader make sense of such strange sights she made a series of comparisons between the animalcules and other objects and animals, real and fairy tale: they were 'rare and clustering gems', 'lilies...with silver stems', 'serpent-forms' gliding through 'banks of moss'; or 'fairy bells', 'Ringing their chimes in fancy's ear'.[171] For Kribble-Krabble animalcules were revolting people; for Mantell, nymphs; for Catlow, gems and flowers and fairy bells: returning these odd entities to the familiar realm through metaphor and simile was one way in which to make sense of them, and to teach their similarities and differences. For instance, the *Paramecium* was colloquially known as 'the slipper'. Just as the familiar water was made strange, then, the strange animalcules were made familiar. Both types of unfamiliarity were, however, brought together in an 1842 French image, in which the animalcules themselves viewed a painting ('toile') of the animalcules found in a 'goutte d'eau', a drop of water, as seen through the oxy-hydrogen microscope ('le microscope solaire'). In a return to the combative origins of this section, the creatures were shown during *une bataille de ces larves*': a drop-bound battle. Not so far away from the Revolutionary French streets after all.

A drop of london water

There were also horrified, not edifying, reactions to the enlarged drop of water: being able to see the true face of nature was not necessarily a good thing, especially if what was revealed was a 'Monster Soup' (see Plate 3). Water was a particularly problematic substance at the mid-point of the nineteenth century. These were years of cholera epidemics and successive debates over water purity; the substance became a key focus of research for chemists and medics.[172] Lurid reports reached the ears of wide audiences, providing ample fodder for merely imagined horrors that, some argued, were more powerful than the actual risks of imbibing polluted water. As one wag commented, 'your Stomach would turn because your mind turned first'.[173] Considerations of what lurked in your drinking water, where that water came from and where it went, were crucial. Most famous, of course, was John Snow's identification of the water pump as the source of a cholera outbreak in the 1850s. One response to these expert medical debates was to invoke wonder as well as terror, and to turn such potentially prosaic issues into a fairy tale.

Satirical prints and periodicals had great fun with the water-drop controversy, lampooning the circular microscopic images of contemporary medical journals such as the radical *Lancet* (see Fig. 13) that have been shown to have had a significant impact on later fairy painting.[174] One image from the comic periodical *Punch* exploited its audience's familiarity with bounded magnified perspectives to portray 'The Wonders of a London Water Drop': rather than the positive associations of glimpsing this hitherto unnoticed world, however, *Punch* wryly inverted the contemporary representations of lurking organisms which caused disease.[175] Instead of pathogenic animalcules they became chimerical pathogenic aldermen, arguably the real cause of the sanitary problems. *Punch* also connected, in its spoof

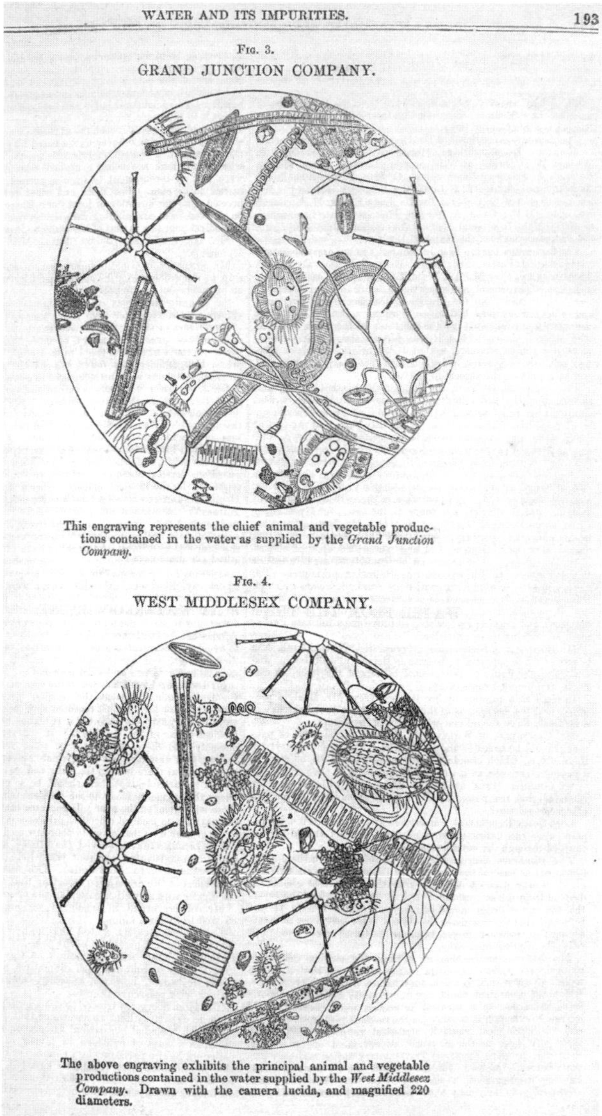
WATER AND ITS IMPURITIES.
193
Fig. 3.
GRAND JUNCTION COMPANY.
This engraving represents the chief animal and vegetable productions contained in the water as supplied by the *Grand Junction Company.*
Fig. 4.
WEST MIDDLESEX COMPANY.
The above engraving exhibits the principal animal and vegetable productions contained in the water supplied by the *West Middlesex Company.* Drawn with the camera lucida, and magnified 220 diameters.

Fig. 13. 'Water and its Impurities', from *The Lancet* (1851). Produced to accompany an expert medical report, these circular images, revealing the previously hidden contents of an invisible drop of water, were often re-created in introductory works, as well as recast for comic effect.

rhyme 'This is the Water that John Drinks' (1849), the stories, rhymes, and lore of childhood with the rancid liquid available in the water-butts on city streets, and even in middle-class homes. The repetitive rhyme, and its illustrations, moved outwards from a homely glass of water, to consider its sources and hidden horrors, revelling in 'the sewer, from cesspool and stink, | that feeds the fish that float in the ink- | y stream of the Thames with its cento of stink, | That supplies the water that JOHN drinks.'[176] In a similar way to the microscopic images, the rhyme destabilized an everyday object, claiming that with an enhanced understanding, here of where the water came from, rather than what it looked like, the familiar, reassuring artefacts of quotidian life were actually repellent. And it (mis-)invoked Coleridge's mariner to lament, 'Water! Water! Everywhere; and not a drop to drink'.

This refrain also lay behind Henry Morley's 'The Water Drops: a Fairy Tale', which appeared in Charles Dickens's periodical *Household Words* in 1850 and was one of the miscellany's many discussions of both scientific and social issues of the day.[177] It introduced the adventures of the Cloud Country people, who dwelled in the skies above the city, to discuss serious political and public problems of polluted water, particularly in poor areas of London. In Morley's story, each particular drop of water or gust of wind became a distinct fairy, a bounded independent character who could converse with River-Drops, or whirl off to play 'pool-billiards with a fleet of ships'.[178] The framing narrative featured a familiar trope: a contest between suitors for the hand of a princess, but was reworked with meteorological overtones. The suitors included Nebulus, Nubis, and Nepelo, subjects of the Prince of Nimbus; the princess was Cirrha, daughter of King Cumulus. For expert readers, the names of these characters (and the supposedly 'bad' and offensive vernacular 'mackerel', 'ball of cotton', and 'cat's tail' alternatives) would act as puns on cloud classification; but for others they would be more like the exotic words found in Dickens's favoured *Arabian Nights*.

The story began by contrasting scientific and fairyland descriptions of a place not in the east, as we might expect from such oriental allusions, but 'far in the west', glimpsed in the hues of summer sunsets. Its appearance, the narrator claimed, was usually 'accounted for' 'by principles of Meteorology', but children had a different explanation: 'it is well known in many nurseries, that the bright land we speak of, is a world inhabited by fairies.'[179] A purely scientific account of the sky's appearance at dusk was deemed neither sufficient nor suitable for the nursery, where it was 'well known' that fairies caused meteorological phenomena. The court of King Cumulus, and the enticing properties of his daughter, were elaborated upon, Cirrha's delicate and pale complexion and deportment suitable attributes for her cloudy incarnation, and desirable counterpoints to the suitors from the land of Nimbus.

From fairyland, the tale descended, along with the various suitors, to the streets and sewers and kettles of London, following the path along which each drop flowed. The suitors were ostensibly endeavouring to do the most 'useful service' in their quest for fair Cirrha's hand: and agreed amongst themselves that the metropolitan conurbation would provide the greatest opportunities for do-gooding. Some had their hopes dashed by the 'lottery' of where they landed, embarking upon brief and rather unimpressive versions of the water cycle:

> One [who] had been the most magniloquent among them all, fell with his pride upon the patched umbrella of an early breakfast woman, and from thence was shaken off into a puddle. He was splashed up presently, mingled with soil, upon the corduroys of a labourer, who stopped for breakfast on his way to work. From thence, evaporating, he returned crestfallen to the Land of Clouds.[180]

Others deliberately dropped into impoverished areas of East London where they believed their help would be invaluable; however, on becoming part of a Bermondsey puddle, Nebulus was horrified by the bathing and cooking that was carried out in poor families with 'foul fetid' liquid, drawn from the 'putrid ditches'. Once on the ground, Nubis, on the other hand, was 'surprised to find, on kissing a few neighbour drops, that their lips tasted inky. This was caused, they said, by chalk pervading the whole river in the proportion of sixteen grains to the gallon.' This led Nubis to reflect on the 'bad business' of polluted water in general; and to note the foreign substances that were already being incorporated into his body: a clear ventriloquism of the authorial viewpoint.

Throughout, the truth of the story was affirmed by bracketed references to the sources of the information spoken by the travelling fairy-drops: intruding on the narrative they pierced the fairy tale veneer and reminded readers they were learning about actual events. For instance, Nubis's conversation in a water-butt was taken from a contemporary medical report:

> 'How many people have to drink out of this butt?' asked Nubis.
> 'Really I cannot tell you,' said a neighbour Drop. 'Once I was in a butt in Bethnal Green, twenty-one inches across, and a foot deep, which was to supply forty-eight families.' (Report of Dr. Gavin.)[181]

The conviction that through fanciful fable concrete and true realities could be imparted, from the corpses of cats floating in 'neglected' metropolitan cesspools to the types of discussions about water and sanitary reform ongoing in well-to-do houses, was central to the success of this story.

Of course, such social commentaries and prevalence of 'monster soup' imagery could also be an exaggeration of the contemporary state of affairs; widespread beliefs in the myriad creatures and many a foreign substance to be found in every drop of water had, one commentator declared, been encouraged by the circulation of this circular image on the page and on the stage:

> Many of these curious beings are tolerably familiar to the public, through the medium of the oxy-hydrogen microscope so popular at exhibitions, and are generally supposed to be inhabitants of every drop of water which we drink. This, however, is not the case, as the water is always prepared for the purpose; hay, leaves, or similar substances being steeped in it for some weeks, and the turbid scrapings placed under the microscope to discompose the public mind...[182]

Just as the water itself was unveiled to have hidden content, so too did this remark unmask the sleight-of-hand that went into the seemingly unrehearsed trick of animalcule revelation. That these were inhabitants in 'every drop of water which we drink' was a crucial rhetorical claim for social reformers to make, though it could lead to disappointment when fresh water was viewed down the domestic microscope.

The combination of water, cleanliness, and domestic fairyland became conjoined in the marketing of new household fairies from the 1860s: rather than the traditional helpful homely creatures such as brownies and pixies, however, these took the very modern form of bars of branded soap. 'Fairy soap' was first advertised in America in the second half of the century, and the 'Household Fairy' trope was now used to emphasize rather the magical powers of new disinfectants and detergents, with the advertising slogan: 'Have you a little "Fairy" in your home?' Fairies were now the means of dispelling germs, not of revealing or conveying them. Soap bubbles themselves were simultaneously an object of expert scientific experiment, a marketing strategy and commercial commodity, and a child's plaything: a magical object. 'Fairy Soap' is now known to almost every British household as 'Fairy Liquid'. But it can still be used to transform plain old familiar water into a frothing rainbow-filled wonderland.

A drop of water on its travels

'How are you to enter the fairy-land of science?' In 1879 Arabella Buckley opened her children's book with this appeal to her audience. In so doing, she revealed questions of paramount importance for Victorian writers eager to recruit children to an active study of nature: what was the most appropriate way of beginning to learn about scientific subjects? Indeed, what were the most appropriate beginners' subjects themselves? Buckley's own answer was definitive: 'there is but one way. Like the knight or peasant in the fairy tales, you must open your eyes. There is no lack of objects, everything around you will tell some history if touched with the fairy wand of imagination ... '.[183] For Buckley, children could learn about science through stories, told with and about the plenitude of illustrative objects at hand in the Victorian home and garden: a piece of coal, a primrose, a sunbeam, a bee, or a drop of water. In this structure, Buckley followed the popular 'object lesson' format of the nine-teenth century, which took a particular everyday artefact as its origin: Faraday's Royal Institution lectures on the 'Chemical History of a Candle', or Huxley's lecture 'On a Piece of Chalk' are other examples.

The fourth of ten lectures delivered in the spring of 1878 to an audience of children in St John's Wood, London, and compiled in book-form as *The Fairy-Land of Science* the next year, 'A Drop of Water on its Travels' asked Buckley's audience to 'spend an hour' 'following' around a 'glistening drop' lifted by finger-tip from a basin of water.[184] The children were encouraged to ask such questions as 'where this drop has been?', or 'what changes has it undergone, and what work it has been doing?' Unlike the animalcule- or pollution-based stories, these introductions relied instead on the connections between clouds and kettles, steam power and the sea, frost and forces, encountered and elaborated upon during the water-drop's biography.

Though they appeared as embodied creatures on the cover of her first edition (see Plate 4), for Buckley fairies were rather to be equated with the invisible forces of nature: cohesion and crystallization, for instance, worked together in freezing water-drops into snowflakes. Thus, Buckley's fairyland references were used as framing devices, entry- and exit-points, just as she hoped the wider fairyland of science would for the world of science as a whole. This is how she managed the transitions from actual observation to general principle; from nature to fantasy; from garden to fairyland. These transitions did not involve large changes: as Buckley had explained in her first lecture, the closeness of her audience to the subjects they were to be taught was paramount:

> this land is not some distant country to which we can never hope to travel. It is here in the midst of us, only our eyes must be opened or we cannot see it.[185]

Many children were actually experimenting with water in these years: a fictionalized family that appeared in *Household Words* in 1850, for instance, were introduced to 'The Mysteries of a Tea-Kettle' as one of a series of 'comical chemical' after-dinner conversations. Importantly, in this presentation it was the boy, rather than his more aged relatives, who was the fount of knowledge about the kettle and its watery contents. He had been to the lectures at the Royal Institution, famously still given every Christmas, and, his despairing mother lamented, 'had been full of it ever since'. For the children of Astronomer Royal John Herschel, too, water was one of the subjects to which elder sibling Bella could introduce her younger charges:

> I want to know all about air! & about water & steam & heat, & weight, & pressure, & resistance & attraction &c &c &c ... There's a modest little want for you! The immediate reason, is that one day in talking to the school-children I plunged into the subject of clouds & dew & then found that I could tell them

a great deal that they could quite understand about the atmosphere, the properties of air, & of water—with ample references
of course to tea-kettles, roasted chestnuts &c. —but still I feel
on rather unsafe ground about it all...[186]

The water-drop was one way in which children could be
'plunged' into such subjects as 'clouds & dew', 'tea-kettles,
roasted chestnuts &c.', particularly by following the drop
around its global journey.

In an 1870 work, the water-drop would come to life itself
to narrate its own history.[187] This, one of Annie Carey's
series of autobiographies of such domestic quotidiana as
salt, coal, and 'a bit of old iron', used a first-person narration
to speak directly to a group of four differently aged children.
Using dreadful puns on the character and voice of the water
(in particular its 'liquid tones'), Carey displaced the authority
of the children's author from her own role, as children
arguably learned straight from nature itself. Children could
now discuss scientific topics with natural objects directly:
they could hear an unmediated, authoritative 'voice from
nature', rather than nature's proxy in the form of a
teacher.[188] An isolated and bounded fragment of nature
was in this way given a particular character; and Carey
could demonstrate that the commonplace and the marvellous were part of one larger whole: the familiar could be
used to access the unfamiliar through such synecdoche. As a
conversation with the grain of salt made clear, this type of

scientific education could be achieved without supernatural aid, instead occurring through sensory training:

> This grain, as Edith looked at it, appeared to grow gradually larger and larger, till at last Lilly exclaimed, 'Oh look! look, Edith! some fairy must have touched it and made it all at once so pretty and smooth, just like a small glass box, only I do not see where to open it.'
>
> 'No, little Lilly,' said a brisk, clear voice from out of the middle of the box, as the child called it, 'no fairy has touched me; I am just what I was before, a grain of common salt; it is your eyes that are touched; and you see me more correctly.'[189]

In these ways, Carey's work explicitly compared fairy tales to her introductory works, as alternative sources of wonder: the first story, for instance, had opened with her group of 'children, sitting round the fire one cold winter's afternoon'. One child, Arthur, asked for a fairy tale; another for her elder sister to 'break up that ugly, dark lump of Coal on the top of the fire'. 'Make it blaze,' she requested: 'I like to see the flames jumping about; they always seem to be alive.' As if responding to her desire for animation, the lump of coal interrupted, and said that he would tell his 'own history': a story 'as wonderful as any fairy tale can be'.[190] An advertisement for the book in *The Times* explicitly categorized it as part of the fairy tale genre: *Autobiographies* were 'a series of fairy tales, in an entertaining form, to teach the child simple, scientific knowledge, that will be useful both now and in after life.'[191] The *Athenaeum* also invoked fairy tales, terming 'Miss Carey's autobiographies' 'delightful', both for the correctness of the facts she imparted, as well as the 'graceful lightness and vivacity' with which they were told, 'which makes them as entertaining as fairy tales'. More importantly, the facts of the sciences themselves were deemed so powerful—and so correctly presented—as to override any misgivings the reviewer might have about the choice of fictional presentation: 'We do not, as a general rule, approve of the plan of turning the acquisition of useful knowledge into a mere amusement; but the elementary facts of natural science are so fascinating and so wonderful, that when put before either children or grown persons with any sort of skill and power of narration, they cannot help being attractive; and Miss Annie Carey has the gift of being able to do justice to her subjects.' The 'Drop of Water', the reviewer confessed, was her 'favourite'.[192]

Down the microscope

Agnes Catlow's decision to take her readers through the magnifying-class—travelling down 'the brazen tube' into a novel wonderland—was to be echoed in an early twentieth-century work that also put an imaginative journey into microscopic dimensions to work. 'Alice Down the Micro-scope' (1927) was a retelling of Alice's adventures written by University of Cambridge biochemists and published in their magazine, *Brighter Biochemistry*, which ran from 1923 to 1931.

The tale brought together the optical technologies they used every day with their favoured childhood stories, the bathetic combination of jargon and juvenilia typical of the sophomore humour of such types of publications.[193] Ever since its inception, Carroll's *Alice* had been a very influential commentary on the relationships between scientific reasoning and fantastical imagining, didacticism and childish curiosity, the natural world and its peculiarities, and the strangeness lurking just beneath the surface of (or on the other side of) the quotidian. First presented with a more geological twist as *Alice's Adventures Underground*, the complex connections and scientific specificity of a charming story told by an Oxford logic professor one sunny afternoon have been teased out by many scholars. Whether the natural history of *Wonderland's* animals, its implicit discussions of new mathematics that dealt with higher dimensions, its explicit reworkings of Isaac Watts's didactic verse, or its immortal incarnation of a very curious child, *Wonderland* and its looking-glass sequel have been mined for their relevance to specifically Victorian debates.[194]

'Alice Down the Microscope' was a short journalistic tale, presented as a vision that Alice experienced after availing herself perhaps too liberally of the 'audit ale' at lunch at her Cambridge college.[195] Bored with 'looking down the microscope trying to see all the things her Professor had told her she ought to see', and finding only 'vaseline and air bubbles' rather than 'motile bacteria', Alice expressed doubt they were 'there at all', and wished she 'could go down and see':

> As she said these words her stool gave a violent jerk and she was shot into the air and alighted a second later on the eyepiece of the microscope. Don't ask me to tell you how she grew very small (3 x 1.5μ), nor yet how she found her way through the various lenses of her microscope till she splashed into the immersion oil; when you are older and understand more about permeability and lattice structure these things will be quite clear to you.[196]

Adopting the tone of a Victorian professor—reminiscent, perhaps, of Charles Kingsley's *Water Babies* as well as Carroll's *Wonderland*—the narrator mixed both technical detail and avuncular hectoring.[197] Just as in the original story, Alice encountered an inhabitant of the strange world she found herself transported into, only rather than dodos and caterpillars this time it was 'a sausage-shaped person somewhat larger than herself, whose whip-like tail was still twisted round her legs'. This 'person' introduced himself as 'Bac. Pyocyaneus', or '"Pyo" to real friends'. An illustration depicted Alice and Pyo together, as he proceeded to take her on a tour of the environs (see Fig. 14). This served to satirize the educational institutions themselves, for instance when the duo found themselves in 'a large lecture theatre, where an active little coccus was delivering a lecture on "Housing Conditions in the Human Body"'. The lecture discussed the 'clearance of the slum areas in the appendix and large intestine', and discussed the potential for further 'colonization' of

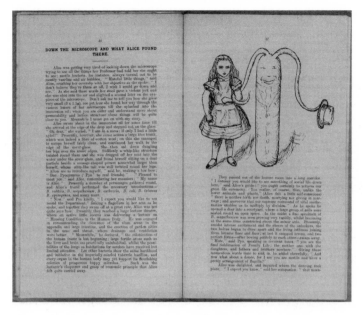

Fig. 14. 'Down the Microscope and What Alice Found There', from *Brighter Biochemistry* (1927). A recognizable Alice, familiar from John Tenniel's original illustrations to accompany Macmillan's 1865 edition of *Alice in Wonderland*, links 'flagellae' with a bacterium, down to whose scale she has been shrunk. While the typical shape of the bacterium has been retained, an attempt to provide a suitably anthropomorphic companion for Alice has been made with the humorous addition of a suit, tie, and trilby, as well as a ghostly hint at a face.

the liver, brain, and lungs: in these ways, it sent up contemporary debates over hygiene, class, and empire. Alice also watched live cell division in action, what her bacterial companion called 'the final Sublimation of Family Life'; as an evidently 'motile' bacterium (with, furthermore, a 'pretty arrangement of flagella'), she was then invited to dance, and in the 'pitiable efforts of non-motile forms...recognised at once the Bacterial Origin of the Charleston'.[198] Other, more familiar scenarios from the Victorian originals were recast, for instance when Alice is put on trial, accused of being a pathogenic member of the Human Race and having committed treason against 'the Bacterial State'. 'In vain Alice pleaded that she was only a poor student incapable of writing scientific articles.' As part of the trial, Alice was subjected to the same kinds of process the students engaged in with their scientific specimens every day:

> Alice spent an awful week; she was stained and counter-stained and decolorized and washed in acid and alcohol and put up in Canada balsam. She was shifted from sugar to sugar and invited to liquefy gelatine; no wonder she bored repeatedly.[199]

At the outcome of the trial, Alice was found guilty, and woken by the Professor who found her current preparation 'in a dreadful mess'.[200]

Well into the twentieth century, then, these wet wonder-lands proved there was more to drops of water than met the eye. They could be revealed as the home for marvellous creatures or dancing bacteria, a miniaturized city-in-revolt, an invisible realm to be journeyed to, where one was acted on by fairy forces, or as distinct travelling fairy creatures. These were just some of the ways in which introductory writers used a droplet of water to present imaginative reworkings of microscopic images and debates over water purity and powers for their readers. In particular these stories show how fairies and fairy tales were used to explore the unseen world just beyond our senses. This world could be a connection to mythical realms of nymphs and dryads; a comforting place of magical powers as in Buckley's benign fairy forces; or a threatening home of horrors, a 'monster soup' of disease and danger. In particular, a focus on water elucidates how the familiar world was unveiled before those with scientific understanding, training, and instruments. But looking closely at the everyday world was not the only way to enter the fairyland of science: perhaps, as we shall go on to investigate in the next chapter, the story of how that everyday world had itself evolved was the true fairy tale of nature.

4

Wonderlands of Evolution

'Ah! what if we're all in a book,
And just being written about?'
May Kendall, 'A Pious Opinion'
Dreams to Sell (1887)[201]

Smelfungus Dryasdust was troubled by a fairy tale. 'One can readily understand', the imagined correspondent wrote to *Punch* magazine in 1879, 'the pumpkin changed into a carriage, the rats into footmen, and the other arrangements for the Transformation Scene wrought by *Cinderella's* scientific godmother'. These were 'evidently a mythic foreshadowing of some great Darwinian Doctrine of Evolution', and therefore 'reasonable enough'. 'But a slipper of *glass*! The thing is preposterous.'[202] Taking issue with the linguistic confusion which had, by the late 1870s, converted the orphan's footwear from *vair* to *verre*, or fur to glass, Dryasdust demanded accuracy of his fantasy.[203] He went on to 'let

110

loose' 'scholarly criticism on the nursery', giving plausible explanations for anything from the particular 'fast-growing plant' which grew from Jack's magic beans (not a beanstalk, but a Eucalyptus, he suggested) to the unusual colour of Blue Beard's facial hair (a hairdresser's fancy for which 'documentary evidence may yet be forthcoming in the Archives of Brittany').[204] This was framed, by Dryasdust and by *Punch*, as an act of iconoclasm: daring to apply the latest scientific knowledge, as well as new techniques of biblical analysis and close-reading, to the sacred fairy tale canon.

However, highlighting the close relationship between Darwin's theory and fairy tales was not uncommon in the second half of the nineteenth century: several children's authors and their works drew heavily on fairy tale forms and allusions when dealing with evolutionary material.[205] From nursery classics (*The Water-Babies* by Charles Kingsley, 1863), to introductory science books (*Life and her Children* by Arabella Buckley, 1880), critiques of some aspects of evolutionary theory (*The Wonderland of Evolution* by Albert and George Gresswell, 1884), not to mention treatments of Darwin's own life, fairy tales were—it transpired—a particularly useful means through which to talk about one of the most talked about topics of Victorian Britain, especially when addressing young audiences.[206] Timeless questions over the history and future of mankind—addressed in folklore as much as in scientific enquiry—gained a new potency in an evolutionary age: man's, and the fairies', place in nature, as well as the place of fairy tales in nature-writing, were up for grabs, and could be conceived of as questions of family relations. The particular form of the fairy tale, therefore, was more than just a familiar framework that attenuated potentially explosive scientific context; it made its own contribution to how scientific theories, scientific practice, and disciplinary boundaries could be thought of, and to the relationships between the natural and the supernatural worlds. Evolutionary theories provoked particular questions of faith and religion, which were dealt with in various ways by children's authors, in line with their own commitments, be it to spiritualism, Anglicanism, or a designing Divine intelligence. Old religious conceptions of the book of Nature were reworked with new, fairy tale plots, as who might have written those stories was brought into question.

Evolution in the family

Several books for children introducing evolutionary theories were published in the second half of the nineteenth century. Some, such as Wendell Phillips Garrison's *What Mr Darwin Saw* (1879), lauded Darwin himself as an expert observer, recasting material from the *Beagle* voyage as an introductory natural history book, or focused on telling Darwin's life-story as heroic exemplar.[207] Others, such as two books by Arabella Buckley, whom we have met as the author of the *Fairy-land of Science*, gave more detailed introductions to what has been called the 'evolutionary epic': *Life and her Children* and *Winners in Life's Race* (1883) pieced together a sweeping story that encouraged young readers to think of themselves as part of a bigger, multi-species family, incorporating all past and present creatures on the planet.[208] Of these evolutionarily inspired works, none, however, has had such longevity or inspired such scrutiny as Charles Kingsley's *The Water-Babies* (see Fig. 15). Tying together physical, moral, social, and spiritual evolutionary themes, the nursery classic made pointed commentary on scientific debates of the day.[209] In different ways to Buckley, Kingsley also championed a particular type of imaginatively informed scientific investigation, and examined familial relationships and responsibilities as well as the metamorphosis of species and society. Perhaps the most explicit way in which Kingsley dealt with the familial relationship between men and the rest of the natural world was with his portrayal of the degenerated race of the Doasyoulikes, whose regression backwards through the stages of human evolution revealed how closely man was related to the apes.[210]

Kingsley's tale held up many familiar themes that we have already encountered. With his story of the poor chimney-sweep Tom, who was converted into a water-baby when falling into a river, he attacked the 'Cousin Cramchild's

Conversations' and 'Aunt Agitate's Arguments' that sought to foist factual matter alone onto their young audiences, Cramchild's name evidently inspired by Dickens's M'Choakumchild.[211] Rather than such stilted means of conveying accurate scientific information, which by the 1860s were indeed rather out of fashion, Kingsley instead wove together observations on natural history and evolutionary theory as the scientific warp to a societal weft in his moral fable, which chronicled Tom's series of evolving adventures. Kingsley's narrator had a strong identity, and often broke through to speak directly to his young charges, cautioning them, for instance, against the people 'who think there are no fairies', and counselling that the 'most wonderful and the strongest things in the world', from the steam of the steam engine to life itself, are those which 'no one can see'. Fairies, after all, may be 'what makes the world go round': the invisible water vapour driving the engine of the universe.[212] Kingsley's narrator also guided his readers on how to combat those opponents of the more flexible system of nature, with its metamorphosing chimney-sweeps, presented in the text:

> If he says that it is too strange a transformation for a land-baby to turn into a water-baby, ask him if he has ever heard of the transformation of Syllis, or the Siatomas, or the common jelly-fish ... Ask him if he knows about all this; and if he does not, tell him to go and look for himself; and advise him (very respectfully, of course) to settle no more what strange things

Plate 1. George Baxter's coloured print of the Crystal Palace grounds shows visitors dwarfed by the exaggerated model monsters, the impressive fountains, and the glass structure itself.

Plate 3. 'Monster Soup commonly called Thames Water', William Heath (1828). This often-reproduced image depicts an early representation of the revulsion audiences felt upon viewing microscopic perspectives on drops of water. Such circular perspectives, and characterization of minute organisms as forming a 'monster soup', were arguably an influence on later fairy paintings.

SIDNEY'S INTRODUCTION TO THE FAIRY

Plate 2. 'Sidney's introduction to the fairy', from *Fairy Know-a-Bit*. The diminutive Fairy Know-a-Bit, dressed as a tiny lecturer, meets the boys he will teach. The fairy is shown in the library which he has made his home, reinforcing his learned persona.

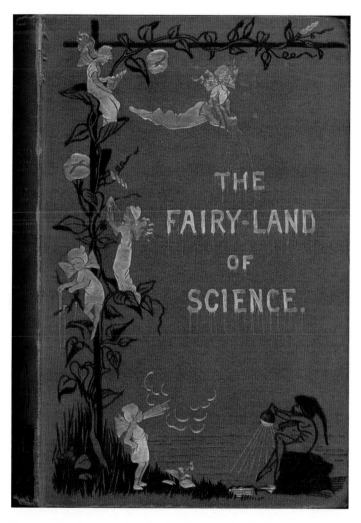

Plate 4. The cover of Arabella Buckley, *The Fairyland of Science* (1879). Whereas in the text of Buckley's book her fairies appeared as invisible forces, on its cover their gilt forms more closely resembled pixies, as they were shown interacting with natural objects such as shells.

A FLIGHT TO THE MOON.

Plate 5. 'A Flight to the Moon', from Agnes Giberne, *Among the Stars: Or, Wonderful Things in the Sky* (1885). Unlike some of the more domestic illustrations in Giberne's work, this allegorical image depicts an angelic 'flight to the moon', accompanying an imaginative journey in the text. In both Giberne's images and words, everyday and more fantastical ways of learning about astronomy were combined.

Plate 6. 'A Ride With the Sun', from Lizzie W. Champney, *In the Sky-Garden* (1877). Just one of the wonderful illustrations that Lizzie Champney's husband, James Wells Champney, or 'Champ', drew to accompany her introductory astronomical tale. The illustrations mined classical myth and lore, to complement the new 'fables' of astronomy contained in the text. Here, the heroine of the text can be glimpsed riding in Helios's chariot. The innovative typography is also typical of these illustrations.

THE WISHING-CARPET.

Plates 7 and 8. 'The Wishing-Carpet' and 'The Electric Boots', from Forbes E. Winslow, *Fairy Geography*. In these two illustrations, the novel perspective granted by the balloon is put to use in depicting the view from a 'wishing-carpet', an updated version of the magic carpet, and also whilst travelling through the air wearing 'electric boots'. Such a bird's-eye view, with rivers and towns, mountains and bridges, far below, was invaluable for teaching geography, complementing other illustrations in the text that showed particular earthbound sites and places.

THE ELECTRIC BOOTS.

Plate 9. The cover of L. Frank Baum, *The Master Key* (1901). This depicts a key moment in Baum's electrical fairy tale: blinding light that signified the appearance of the Demon of Electricity, almost like the summoning of the genie from Aladdin's lamp.

Fig. 15. 'Professor Owen and T. H. Huxley examine a water-baby' (1886). Two men of science—recognizable as Richard Owen and Thomas Henry Huxley—are shown peering rather sceptically at the bottled protagonist of Charles Kingsley's *Water-Babies*.

cannot happen, till he has seen what strange things do happen every day.[213]

This passage is characteristic of the way in which the narrator's avuncular persona enabled much learning to be lightly worn, with scientific references (for instance to current debates about the structure of the humanoid and ape brain) that may well have not been intended for the young readers themselves but for their parents, intermingled with advice on proper conduct ('very respectfully, of course'), as well as pointers on the best way to become an authority on what was possible in the world: looking at things for oneself, but without prejudice, and with imagination, just like the best men of science.

As the depiction of the 'fairy Science' revealed, a maternal emphasis was found throughout Kingsley's work: he brought in what one scholar has called the 'Mother Nature figures' of Mother Carey (the source of all life), Mrs Bedonebyasyoudid, and Mrs Doasyouwouldbedoneby, figures who resembled the female instructresses of the early century, and who were certainly also related to Kingsley's Madam How and Lady Why, stars of his later book which also dealt with how to learn about natural history.[214] With connotations of Mother Goose, such characters also alluded to the very origins of the fairy tale genre with Charles Perrault's collections, as well as highlighting the motherly role as story-teller to a family of children. Both Buckley and Kingsley brought evolution into the family, with Mother Nature, 'Life', or various female instructors, cast as its benevolent head.[215] But for other authors it was fairy godmothers, rather than Mother Nature, who provided the best way to introduce debates over evolutionary theories to new audiences, and to reflect on the role of both lay and divine authorship itself (see Fig. 16).

Evolutionary authorship

Something of the *Water-Babies*' message, of a moral and spiritual, as well as a physiological, evolution from benthic to human organism was echoed in the Gresswell brothers'

Fig. 16. The original sketch for the 'Science Fairy', by Edward Linley Sambourne, personified many recent scientific achievements in allegorical form, from a steam-based corset to clothing covered in geometrical shapes and the names of specific disciplines, and an electrical beacon. A mortar board on the head of 'Science' emphasized her scholarly status.

1884 book for children, which transported their readers to a *Wonderland of Evolution*.[216] Just one of the post-*Alice* books which capitalized on the popularity of all things 'wonderland', the book's narrative was, however, less picaresque and more progressive: its first-person narrator transformed into an evolutionary catalogue of appropriate organisms, beginning with the 'lowly form' that represented the origins of life on earth. In a more overtly Carrollian trope, the narrator in his varied guises was able to converse with the lowest to the highest forms of life on earth.[217] There had been literary precedents for this type of narrative choice, from the zoological 'Transmigrations of Indur' in the late eighteenth-century children's miscellany *Evenings at Home* to John Mill's spirited fossils of 1854, which we encountered in Chapter 1; most famous, and most transmutationary, had been Alton Locke's dream sequence, which evolved Kingsley's eponymous protagonist.[218] But rather than a simple endorsement and presentation of evolutionary ideas, in a narrative form deemed suitable for juvenile audiences, Albert and George Gresswell made a more subtle argument. The brothers' usual territory was veterinarian, their expertise more suited to oxen, sheep, horses, and cattle, rather than protoplasma, fish, or lizards; on turning to evolutionary theories, they attempted to 'illustrate the subject in as interesting a manner as its intricacy will allow'.[219] For Albert and George, the pages of a children's book was an appropriate venue for disentangling that 'intricacy' of evolutionary ideology: they would pick apart its threads, agreeing that development of the earth's organisms had taken place, but using fairy guides and talking animals to save the phenomenon of divine power, and downplay the role of chance. Even, therefore, in such apparently superficial and fanciful juvenile tours through the history of life, precise interventions in specific scholarly debates were made. The story of evolution was not just one of Darwinian fame, but also of those other nineteenth-century figures such as Ernst Haeckel and Herbert Spencer who put an evolutionary agenda at the heart of their work.[220]

The fairy element to the Gresswells' book was introduced most explicitly through the characters of 'the fairy Chance, an airy magic sprite', and her 'friend, that cunning sylph, Evolution'.[221] In the book's opening pages, Chance spoke to the lowly narrator, introducing the fairy-godmother role that Evolution would play, to 'perfect and mature all that I initiate'. The powers of scientific observation and analysis were granted in this text not through painstaking labour, but as a supernatural benefaction, akin to the more usual fairy tale christening gifts: 'I now give you also the power of seeing the most minute particles, of hearing all varieties of sound, and of interpreting the murmurs of the very tiniest of my creatures.'[222] With such powers, the audience, along with the evolving narrator, encountered 'endless marvels, inexpressibly sublime', that charted the development of life on earth, from most simple to most advanced organisms.[223]

However, from the outset, the Gresswells' desire to undermine the role of Chance in evolutionary processes was made clear:

> Though deeply impressed with her words, inwardly I felt that, false as fair, Chance could not fulfil her promise; and though ready to follow the steps by which the two fairies were to accomplish each her part of the great work prophesied, I reserved my decision as to a belief in their power to effect this wondrous end.[224]

Such a sceptical, even sarcastic, tone pervaded the later set-pieces in the book, for instance when discussing Darwin's later theories of sexual selection, 'that the gentlemen became beautiful because the ladies were fastidious, and the reason the ladies were fastidious was because the gentlemen became beautiful'.[225] Encounters and conversations with personified animal characters permitted analysis of what was 'true' and what 'exaggeration' in their 'mirthful stories'; such a discussion, of course, mimicked that which the authors hoped their own readers would have, taking apart aspects of evolutionary theory, and assessing their merits.[226] For example, after one such conversation with 'Mr. Lizard', the narrator reflected as follows:

> 'I quite admit,' said I to my friend, as we hastened along, 'that there is an element of truth in what Mr. Lizard has told us, yet there is much exaggeration in his mirthful stories; and if we do not allow that *all* is true, the question still remains, who gave such discriminating power to these brilliantly-coloured

creatures? It certainly was not Chance.' "'No", replied he; "for all chance is direction which thou canst not see."[227]

The archaic language ('thou canst') connoted timelessness as well as biblical cadence; God as divine author, it was implied, was the invisible 'director' one cannot see, who was ultimately driving the evolutionary process.

At another point in the story, the narrator was cast as 'arbitrator' in 'a lively discussion' between 'ants respecting the origin of their society and the question of the division of labour'.[228] The motion in question was the perennial concern of *The Wonderland of Evolution*, continually addressed by its narrative, alongside its didactic purpose of introducing the different forms and conditions of various creatures in an evolutionary sequence:

> The argument of one was that the society had originated by the fostering care of a Designing Power; while the other maintained that all had been the work of far-seeing Chance and good-natured Evolution.[229]

With 'twelve neuters' (suitably impartial, naturally) selected as 'jurors', the narrator proceeded to state his case:

> My friends, in the course of my travels through other lands, I have heard equally absurd arguments advanced by some people; but nevertheless you may be assured that a Designing Hand is clearly indicated. Depend upon it, my friends, a much higher power than the fairies mentioned has been at work. Chance, always acting blindly, as she does even though aided by Natural Selection, could never have initiated—much less perfected—such instincts as those of the honey-bee.[230]

The intricate 'instincts' of the bee were held up as evidence for an authorial guide capable of such refined work; an opinion which was received, the narrator boasted, with an 'almost unanimous decision' 'in favour'. As in Buckley's works—both in *Fairy-land of Science* and *Life and her Children*—for the Gresswells the place of the fairies in the explanation of scientific processes was to stand in between the natural world and the divine 'higher power' that was its originating influence. As preternatural creatures, fairies were well suited for such work, their wings connecting them to angels as well as to insects. As narrative agents, too, these fairy godmothers similarly mediated between the role of character in plots and a role more akin to authorial figures, who could, with a wave of a wand or pen, set events in motion. Most pertinently to an evolutionary context, they could also effect remarkable metamorphoses, from rat to footman, pumpkin to carriage.[231]

Though this regard for a divine presence was felt throughout *The Wonderland of Evolution*, the work also pleaded for the tolerance of differences of opinion, especially with regards to the respective roles of Evolution and Chance. Some characters agreed with the narrator that it seemed 'impossible that the two fairies could have alone worked such wonders', but others had a more violent reaction. Most dramatically, in the case of the eel:

Several obstinate members of this band of electric fishes were, however, so angry with my views that they conspired together to electrify me on the spot. Perceiving their intention, I swam off rapidly, and was soon at a respectable distance, closely pursued, however, by those who would not look at the question from an impartial aspect. They could not understand that every great problem can be looked at from more than one point of view, and were ready to devour all who disagreed with them.[232]

The Gresswells reflected on the occasionally vicious debates over evolutionary theory that were ongoing in lecture halls, periodical columns, and middle-class parlours across the country, as well as pre-empting criticism that their own work could receive. Reception for the work was indeed mixed, if not quite as shocking as the characters' electric experience. For instance, for the *Glasgow Herald*, the *Wonderland of Evolution* was not a successful children's book; in particular, it did not live up to its Carrollian namesake. Though the periodical praised the printing, paper, and typography of the book—crediting the publishers with a handsome addition to their stable, a credit to 'artistic bookmaking'—this was 'all that [could] be said in its favour'.[233] Assuming that the authors had attempted to write something 'in the spirit of Lewis Carroll', the reviewer claimed to have had little to go on save the title: 'there is little to justify our supposition beyond the name and a painful effort now and then apparent to be grotesque, as when we are introduced to a party of molluscs drinking tea and brandy'.[234] The supposedly comical cockles to which the reviewer alluded do indeed read as an unfortunate reworking of that famous tea-party:

> 'Boiled!' screamed every one again at the top of his voice, more vehemently than before, while many of the poor cockles present made a hasty retreat, spilling tea, coffee, eggs, and butter on the floor; 'do you mean to say you want to eat boiled cockles?'[235]

As the Herald reviewer wrote: 'We fancy the authors mean to be funny, but we only find them silly.'[236] It was not just the taking of the name Wonderland in vain that had troubled the reviewer, however: more serious charges were laid at the Gresswells' door:

> They seem to think that all who adopt the theory of evolution … are either atheists or agnostics. It is not at all evident that the authors understand themselves what is the theory of evolution, but that really is of very little consequence. Their attempt to make fun of it is certainly a distinct failure.[237]

The book was therefore deemed not just a failure as a children's book, but also as a scientific primer and a moral guide: as we have seen elsewhere, the accuracy of the knowledge included in these more fanciful reworkings was just as crucial as their fantastical portrayals of talking protoplasm, debating ants, or tea-drinking cockles. The reviewer attacked the writers for what they had been at pains to avoid: a denial of the 'intricacy' of evolutionary debates, and instead an easy conflation of evolution and atheism.

Other reviews of the book also reflected more generally on the speedy adoption in many quarters of evolutionary doctrines, noting 'the haste with which many persons who considered themselves votaries of science and others who put forward no such claims readily accepted the theories propounded by Darwin in his famous work on the origin of species'.[238] Darwin, the *Morning Post* reviewer noted, had 'laboriously collected' the evidence for his theory over two decades; 'yet persons who had not given as many hours to scientific study were ready, after the most hurried perusal of the volume in question, to assent to the principles therein laid down.'[239] Such easily convinced audiences, the reviewer seemed to imply, were swayed by scientific fashion, novelty-seekers eagerly embracing 'the latest thing'. The Gresswells, the reviewer approvingly noted, were not amongst those who had uncritically adopted Darwinian doctrines: they 'refuse, in common with a large proportion of educated and thinking humanity, to regard the evolution theory as a final explanation of the origin of the universe'. That final question of origins, the reviewer claimed, was 'where science fails':

> She cannot tell us whence the original matter from whence all things are compounded came into existence. By chance! Where is the proof? .... He who contends that matter came into existence by Chance (with a capital C) must indeed have the strongest faith ... the authors have succeeded in the task they set themselves, not to confound opponents, but to bring before their readers the urgent necessity of exercising the utmost

caution in accepting the statements of experimental and theoretical science, when it attacks the basis of religious faith.[240]

The conclusion of *The Wonderland of Evolution* signified a marked change in tone, breaking from the fairy tale framework to adopt a lecturing style: these explicit comments were often what had been picked up on in the book's reviews. As one historian of science has put it, 'the Gresswells speak directly to their young audience rather than through their fantastic fairyland' creatures; this permits them to 'warn their young readers that materialism leads directly to the destruction of ethics'.[241] Breaking character, they emphasized that evolution could only be 'the method of procedure of an Almighty Designer'.[242] In so doing, they drew attention to their own role as authors, capable, unlike God, of speaking directly to their congregation about how these things had been created, rather than through myth and fable, exploration and discussion. By analogy, they implied, a thinking brain and a writing hand lay behind the created world: God was the ultimate author of the fairy tale of life on earth. By such self-conscious use of literary devices in the body of the story, and the striking change of voice and mode in the final section, readers could not be unaware of the presence of the authors; and therefore were led to a similar recognition of the indispensable presence of the Divine author. The illusion that the inhabitants of past and present natural worlds can tell us their stories

directly (in the story through conversation and ventriloquism; in reality through scientific investigation) was
shown to be a fictional device, something that did not
happen by chance, but by the actions and intentions of a
particular organism. The authors of this work were the
Gresswell brothers. But the author of the world was God.
And the kind of stories he wrote were fairy tales.

Nature's fairy tale

The fairy tale remained a favoured analogy for telling the
history of life on earth, and for explaining the relationships
between past and present, throughout the nineteenth century. In the December 1868 issue of the periodical *Good
Words for the Young*, for instance, Hugh Macmillan made
explicit reference to the literary work lying under his
readers' feet, and powering their modern world:

> The page of the earth's story-book that tells us the history of
> coal is a very extraordinary one. It is to the familiar appearance
> of the world at the present day what the fairy-story books of
> childhood are to the sober duties and enjoyment of grown-up
> men.[243]

As a juvenile genre, fairy tales, Macmillan argued, were
particularly appropriate for relating the early days of the
planet to audiences in their early days themselves. They, like
the coal strata, captured something precious about a past

age, preserving it until such time as it could be unearthed and put to work. Kingsley would choose to tell the history of life on earth as a 'true fairy tale' in *Madam How and Lady Why* (1870): though he began by claiming his choice of literary form (or, as he put it, 'shape') would ensure young readers 'understand it at all', in fact more was at play: Kingsley reflected that he was, in fact, telling 'the fairy tale of all fairy tales'.[244] The appropriate literary form hence aided a self-reflexive commentary on how—as we have seen claimed by contemporary writers in Chapter 1—folk memories of giant creatures and bloodthirsty 'savages' became 'ridiculous and exaggerated' and 'grew up' into legends of 'fairies, elves, and trolls, and scratlings, and cluricaunes, and ogres'.[245]

The notion of writing nature's own fairy tales continued into the twentieth century: in 1913 children's author Lilian Gask used the form to present the 'earth's story-book' as a fairy tale in her *In the 'Once Upon a Time'*.[246] This extraordinary work was an alternative guide to London's museums: rather than presenting information about the latest fossil discoveries on display in South Kensington as blocks of factual text, Gask reworked material from best-selling geological and anthropological writings alongside guidebooks and images from the Natural History and British Museums. The history of humanity was best expressed, this seemed to say, not through displays of rocks, but through directed plots. Gask subtitled her work, echoing Tennyson, 'a fairy tale of science': as she claimed in its preface, 'not the least penalty of "growing up" is the discovery that we must surrender as untrue the fairy tales we loved in childhood'.[247] With similar rhetoric to, for instance, Arabella Buckley in her *Fairy-land of Science*, she claimed scientific stories were a better replacement for the tales of old. Whereas she identified this genre as a 'new class of fairy tale' 'for the little people of the present day', it was actually a well-established literary strategy by 1913:

> These are the great fairy tales of science which Nature has written for us in the earth that lies beneath our feet, in river gravels, in rocks and stones, and in the depths of the sea itself.
> Beside these, even the imaginings of a Grimm or Hans Andersen seem dull and colourless; and the story of Man long ago, with the curious creatures that preceded him when the world was young, is surely the most wonderful of all.[248]

The superiority of the fairy tales of science was assured, Gask seemed to say; and it was assumed that this particular literary genre was the most appropriate for telling the history of life on earth, Gask's task being 'to unfold a few pages of this fairy tale to the children'. What was most novel about her work was not, however, the fairy tale framing. Rather, it was her anthropological subject-matter: the story of Man, 'the most wonderful of all'.

The protagonist of *In the 'Once Upon a Time'*, Phil, had appeared in her previous work, *In Nature's School* (1908), a story that had introduced the habits and habitats of a range of animals by providing an alternative year of schooling for its unhappily orphaned hero.[249] Its escape from the classroom into more ethological, fieldwork-style modes of learning by living with animal communities evoked and complemented contemporary rhetoric that 'Nature, not books' was the best teacher. This so-called 'Nature Study' movement, endorsed by leading educationalists in universities, schools, and even Scout groups, was then in its heyday across Britain and America, and advocated learning from and with the surrounding environment.[250] The pseudo-sequel to *In Nature's School* likewise began in the school's woods, where Phil encountered a 'Little Professor' at the beginning of the holidays; however, this shift to encountering an expert adult revealed an alteration in didactic strategy from the previous book, whose lessons had been spurred by a communion with Mother Nature. Fairy tales were now, the new literary style and narrative detail seemed to suggest, to be told and retold, rather than experienced directly, as a 'fairy tale that has no ending, and that grows more entrancing as we read on'. The 'Little Professor' showed his Victorian origins with his re-articulation of Buckley's message for a new generation.

From the outset of the story, Phil is presented as longing to visit the metropolitan museums: '"They must be ripping places," he said to a bright-eyed rabbit.'[251] One of the Professor's early explanations—occasioned, as in Nature Study books, by the observation of a particular animal in its habitat—also made museological mention:

> 'Shrews are said to have existed more than two million years ago, for their remains have been found in some of the deepest layers of the earth's surface which have yet been explored. What do you think of that, eh? They are supposed to have been here long before those strange prehistoric monsters whose fossilized bones are now in the Natural History Museum; giant tortoises, and flying dragons, and huge flesh-eating creatures with savage teeth. Man is quite a newcomer by comparison, though most likely he was here as far back as 400,000 years ago.'
>
> 'It sounds like a fairy tale,' said Phil, his merry eyes growing large and round as he tried to picture what the world was then.
> ....
> 'And so it is,' said the Little Professor; 'a fairy tale that has no ending, and that grows more entrancing as we read on.'[252]

Split across a chapter, the placing of this conversation served to reinforce the message that the story of life on earth was never-ending, escaping even the confines of the book format. As we have seen in Chapter 2, the notion that participating in scientific investigation was akin to beginning an unending story can be found throughout these elementary writings. Here, though, the story of mankind was conceived of as a fairy tale, rather than scientific practices themselves. With reference to 'flying dragons' and 'huge flesh-eating creatures', the story—like those discussed in Chapter 1—drew on the gory appeal of monsters, and their consequent conquering by time, to lie in fossilized pieces behind glass, to draw in its young readers, and provide rhetorical counterpoint for the even more impressive story of man.

Gask's book moved through key topics in the history of mankind, from 'The Birth of Knowledge' or 'When Man found Fire' to 'Some Early Hunters'. The locations in which Phil and the Professor's lessons took place often reinforced their didactic content, for instance hiding in a cave mimicked the lifestyle of early man that they were discussing; fire was discussed when the duo needed to kindle warming flames themselves. Phil was also able to make many connections to his own life, remarking how 'We do that sort of thing when we're scouting', bridging past and present practice.[253] Chapters also dealt with such topics as 'Reading the Past', and 'In the British Museum', which gave instruction not just on *what* to learn, but *how* to learn it. Gask gave references to works she had consulted in the writing of her book, including leading scientific works as well as 'several guide books of the Natural History Museum', and 'most interesting lectures in the Galleries of Fossil Birds and Mammals, and Fossil Reptiles'.[254] Other contemporary authors, such as Hutchinson in his *Prehistoric Man and Beast* (1896), also relied heavily on the museum collections, and used them to analyse the relationships between folklore, myth, and the sciences, just as he had done in his *Extinct Monsters*.[255]

We can best think of Gask's book, then, as a reworking and synthesizing of sources, from leading textbooks, to lecture notes, to museum artefacts themselves. Her illustrations, including, for instance, a mammoth and a miner's pick, were credited to 'Blocks' from the Natural History Museum. Her book demonstrated what happened when a plot was added to a museum catalogue, disabusing Phil of his belief from 'some lessons that he has had at school' that 'An-anthropology', or indeed 'anything ending in "ology" must be dull'; rather, it was the source of the best stories.[256] Even when Phil and the Professor visited the Natural History Museum itself towards the end of the story, they could not escape fantastical references, and seeing stories all around them: 'From the end of the hall ran a wide stone staircase which Phil felt convinced must lead to Wonderland.'[257] Another favoured seasonal entertainment, the pantomime, was also compared with the history of the Earth (see Chapter 6 for more on scientific pantomimes), losing out once more to the superior scientific spectacle: 'no pantomime you could ever see', claimed the Professor, 'would give you such transformation scenes'.[258] Indeed, Phil did come to this conclusion at the end of the book:

> After dinner the Little Professor took him to a pantomime, in which many strange beasts came tumbling on to the stage, turning presently to beauteous fairies.
>
> 'It's not a bit funnier than things that really happened!' said Phil, when his laughter would let him speak.

'We're only at the beginning of the fairy tale, you know; and don't forget that you're booked to me for your next holidays.'
'Ripping!' said Phil, falling back as usual on his favourite word.[259]

Museum exhibitions of actual remains could be combined with the language and power of fairy tales to provide the best sources of entertainment, as well as teaching the latest in scientific discoveries.

An evolutionary Märchen

For May Kendall, topics such as evolution provided fertile ground for poetic inspiration. Her 'Ballad of the Ichthyosaurus', for instance, was told from the perspective of a specimen from 'a goodly Museum | Frequented by sages profound': the specimen recalled its previous existence, when 'we dined, as a rule, on each other' and when 'the toughest survived', but bemoaned its poor cranial capacity, and wished, instead, for a 'Brain that is bulgy with learning'.[260] Both Darwin and Owen were name-checked by the poem, as people who were 'cleverly' able to 'restore' this vanished specimen to 'dwell in sweet Bloomsbury's halls'. Kendall's 'The Lower Life', on the other hand, personified Evolution more along the lines of the Gresswell brothers' fairy godmother, as a wish-granting genie, with the poem's narrator, regretting that 'Evolution has not yet | Fulfilled our wishes', wondering why humans have been given 'brains' but not the ability to fly.[261] Evolution was presented as a character with her own agenda, who 'shows no favour' but—like a Carrollian figure—'With knights or pawns pursues the game'. Musing on whether the power of reason or the power of flight be the better benediction, the poem considers the past and potential of an evolutionary sequence from monad to man to, perhaps, an angelic or fairy tale future. The poem is uneasy with the idea that commitment to 'The March of Reason' and the 'important prize' of 'Truth' means that 'We wholly must away with lies', in a vote between 'creed Utilitarian' and the 'flight of airy pinions'.[262]

Both of these poems were contained within the 'Science' subsection of Kendall's 1887 anthology *Dreams to Sell*, alongside odes on trilobites and jelly-fish; the section also included the less obviously scientific 'A Pious Opinion', signed by A.L. and M.K., presumably Kendall and her frequent collaborator Andrew Lang.[263] Its title offered hints of traditional, indeed, devout, musings on morality and the condition of mankind; however, its subtitle ('Mythological') and opening lines immediately undermined a more theological context with direct links to fairy tales, and those who dared to make them the subject of professional study:

> 'CINDERELLA,' they say, 'is not true,
>   'Tis a *Märchen*:'—we listen and wince,—
> ''Tis a Myth of the Dawn or the Few,
>   And she never did marry the Prince ...'[264]

Just like our opening fictional correspondent, Smelfungus Dryasdust, here Lang and Kendall responded to figures such as Hutchinson (the 'they' of its first line), who sought to uncover the folkloric origins of fairy tales and fairies. Rather than revealing the scientific principles behind fairy tale plot-points, as Dryasdust had done, claiming for evolutionary theory the role of fairy godmother, the poets instead declared that such tales were 'not true' at all. The Germanic term 'Märchen' both alluded to the Grimm brothers' famous pioneering work in folklore and linguistics and also reflected a distancing professional vocabulary that could only identify and classify with elaborate, and often multilingual, taxonomic labels. Whereas the poets' protagonists shared a 'wince' at this iconoclastic treatment of an old favourite, their interlocutor continued to dismantle the rest of the tale's untrue aspects: it was merely a 'Myth of the Dawn'; there was to be no happily ever after. The poem's following lines claimed that children must 'leave Cinderella behind', as someone they 'never will find', unlike an actual historical figure such as Henry the Eighth, whom 'they may meet'. Of course, other than in the pages of a novel by H. G. Wells it was equally impossible for a Victorian child to meet Henry the Eighth as it was to meet Cinderella, but the fact that it might *once* have been possible was key to the rhetorical claim. A procession of characters from fiction and fable were then marched out, from Tom Brown to Tom Thumb: for whom, it was suggested, readers will now 'pine', when they are replaced through 'painful but orthodox rigour' with 'the elect of the ages', 'poets and seers', 'singers and sages'. Without writers in prose and poetry who kept these characters alive, the following lines argued, the characters would fade away: "Tis to us for existence they look.'[265]

Nevertheless, it was the authors' *own* existence that they were 'beginning to doubt'. What if we were just characters, being written about in a book? On the one hand, such questions could be dismissed as classical undergraduate philosophical posturing; but it was also a way of thinking about history and prehistory that preoccupied many in the second half of the nineteenth century. What is the difference between history and mythology, between Henry the Eighth and the Seven Dwarfs? As those widespread analogies claimed, life on earth was best represented as like being in a story book. Kendall and Lang took such considerations to their logical conclusions: if we are in a book, then who is writing it? And if we're not being written about at the moment, then we will be in the future, when we have turned from history to prehistory, or to mythology; when our forms have fossilized. The book's opening pages—that fossil record, those folkloric songs and stories—have only survived in part, so 'we must skip the beginning'; 'The end is beyond us indeed', as it is one that 'none but Immortals can read'.[266]

In these ways, like several other works I have discussed in this chapter, Kendall and Lang made explicit the longstanding natural theological analogy of the book of nature, with self-conscious reflection on the relationship between written and real characters. The book of nature was, ultimately, one whose pages were imperfectly preserved, or as yet unwritten, and it had a divine author. Claiming that the fairy tale was the particular type of story-book in which our history was written permitted the incorporation of preternatural creatures who stood in between God and nature in the narrative, like the book's authors themselves. These books went back in time to think about the relationships between the human and divine; but other authors looked to the skies instead, as with optical technologies they searched out and refined the 'fairy story of the heavens'.

5

Through Magic Glasses

Harriet Martineau was not amused by magic lantern shows. As a child, the renowned political economist and writer had been subjected to a series of optical demonstrations, 'on Christmas-day and once or twice in the year besides'. Whereas other children might have oohed and gasped at the strange forms cast onto the parlour walls, in a domestic re-creation of the metropolitan shows we encountered in Chapter 3, Martineau had a more visceral reaction: 'the first apparition always brought on bowel-complaint'. Her 'terror' of 'the white circle on the wall' and 'the moving slides' was just one of a series of juvenile 'panics', what she would term 'a matter of pure sensation', without any 'intellectual justification'.[267] Witnessing the cleaning of the lantern, handling its parts, and 'understanding' its scientific principles was not sufficient to quell her reaction to, for instance, the apparent approach of Minerva and her owl: she could 'remember [her] own shriek'.[268]

Martineau's failure to differentiate between surreal and scientifically produced experiences reveals how close they continued to be, well into the nineteenth century. A longer history of optical trickery was inextricable from new technological developments, from the kaleidoscope to the spectroscope, as instruments, practitioners, authors, and audiences, drew on conventions and associations of natural magic. As the conduit of both didactic lectures and fantastical visions, the magic lantern was a particularly powerful object: put to work in elementary instruction as much as underneath the Victorian stage, using optical principles to conjure modern ghosts.[269] But the magic lantern was just one of what Arabella Buckley, author of the *Fairy-land of Science*, termed 'magic glasses': instruments such as the telescope and microscope which were caught between the wonderful and the everyday, or Martineau's sensational and intellectual.[270] Indeed, authors often combined wonderful and everyday objects and prose in order to teach their young readers about the worlds that were very far away, or very small, realms they could only travel to through these scientific instruments, or on the wings of imagination.

Natural magic

Arabella Buckley addressed *Through Magic Glasses* (1890) to her 'young friends' who had read its prequel, the *Fairyland of*

Science. Just like the elementary experiments she had advocated in that work, she also recommended a practical component to accompany the reading of her latest works, hoping that 'in these days, when moderate-priced instruments...are so easily accessible' 'some eager minds may be thus led to take up one of the branches of science opened out to us by magic glasses'. Even if reading her book was not a prelude to wider study, she took succour from the fact that readers would 'at least understand something of the hitherto unseen world which is now being studied by their help'.[271] Presenting her work not—this time—through her own lecturing persona, Buckley chose a magician as the star of her book, however, the structure of later chapters was more closely reminiscent of *Fairy-land of Science*'s object-based approach, although this time it combined more and less familiar objects in its choice of subjects, some of which could have been drawn straight from fairy tales: 'fairy rings', lichen and mosses, a lava stream, the sun, and Dartmoor ponies. The final chapter, 'The Magician's Dream of Ancient Days', took on a different tone to—like Gask's book—discuss human prehistory, this time in a dream vision rather than a fairy tale. In *Through Magic Glasses*, which opened in the moonlit chamber of a magician, one might perhaps not expect to find such domestic examples foregrounded, but rather a mystical sense of the starry far-off. Nevertheless, just 'down the turret stairs' from that magician's chamber was 'a large science class-room', including instructions on how best to illustrate the phases of the moon, by the light of the magic lantern.[272] Illustrations in the work ranged from its illuminated cover to depictions of children demonstrating basic astronomical principles, to reproductions of recent astronomical images from more expert publications (see Fig. 17).

Buckley followed the earlier example of renowned Scottish man of science, and inventor of the kaleidoscope, David

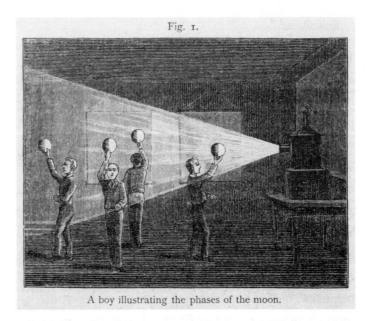

Fig. 1.

A boy illustrating the phases of the moon.

Fig. 17. 'A boy illustrating the phases of the moon', Arabella Buckley, *Through Magic Glasses* (1890). The illustrations in Buckley's *Magic Glasses* varied enormously, from depictions such as this that could readily be enacted by readers at home, to reproductions of recent astronomical photographs.

Brewster in her emphasis on the 'natural magic' of the primarily optical instruments which formed the basis of her introductory essays, which were described in detail in her introduction, their names standing out on the page in capitals.[273] The 'largest magic glass' was named as the 'TELESCOPE', and introduced in the following manner:

> It was only a moderate-sized instrument, about six feet long, mounted on a solid iron pillar firmly fixed to the floor and fitted with...clockwork...yet it looked like a giant as the pale moonlight threw its huge shadow on the wall behind and the roof above.[274]

The spectroscope was described in a slightly different way, as an 'uncanny and mysterious' device, with which the magician could 'read the alphabet of light'; though its purpose was explained, other 'curious prisms' were 'not', the narrator stated, 'to be understood by the uninitiated'. As with her choice of a magician, the language throughout these introductions was reminiscent of arcane tradition, and played on the strangeness of the objects: they were 'uncanny'. The camera, however, was 'comparatively natural and familiar', a 'hooded monk' with 'its tall black covering cloth', rather than a 'giant'; the microscope a 'highly-prized helpmate'.[275] 'If in the stillness of night the telescope was his most cherished servant and familiar friend, the microscope by day opened out to him the fairyland of nature.'[276] In these ways, Buckley introduced the instruments as the key characters of the work: anthropomorphized as characters (giants, monks, servants) of fairy tales, even before further information about that mysterious magician was given (it turned out he was 'Founder and Principal of a large public school for boys of the artisan class'). It was with these magical instruments that the secrets of the universe could be revealed, and even reproduced in the text: for instance, many of the engravings were based on recent astronomical photographs.

The introductory section included a detailed introduction to astronomical phenomena, particularly of the moon: sneaking around at night, the boys from the school were making illicit nocturnal observations. The next chapter, 'Magic Glasses, and How to Use Them', was the keystone to the work, presenting the 'bright and business-like' instruments, diagrams, and similarly 'bright faces' of the magician's pupils, rather than 'ghostly moonlight'.[277] The magician warned his young charges that 'there is no royal road to [his] magician's power': you must 'take trouble', and 'open your eyes and ears, and use your intelligence' in order to develop such scientific skill.[278] A comparison was made between the eye and 'a very simple and pretty experiment' of a magnifying glass, a candle, and a square of white paper, to demonstrate how the optic lenses work: the eye was held up as the most basic of those magic glasses, of course. Throughout her introductions to the microscope and telescope, Buckley made comparisons between the wonders she was introducing and other types of marvels, exclaiming—

after revealing that the light from some stars 'takes 2,000 years to reach us'—at the superiority of the natural world: 'Can any magic tale be more marvellous, or any thought grander, or more sublime than this?'[279] However, it was the spectroscope that was reserved for the final unveiling, and held up as the most marvellous of devices: 'the work of our magic glass, the spectroscope', Buckley revealed, was 'simply to sift the waves of light', to 'tell us what glowing gases have started them on their road. Is not this like magic?'[280] Buckley's magician drew a contrast between the 'true magicians and false magicians' in stories 'of days long gone by', emphasizing that 'the value of the spells you can work with [his] magic glasses depends entirely upon whether you work patiently, accurately, and honestly'.[281]

A closer look at Buckley's chapter on 'fairy rings' reveals how she brought together conventions of the school story with concerns to explain the folkloric objects of the world on scientific principles with a demonstration of introductory microscopy and mycology. The chapter opened with a description of the schoolboys on their 'yearly autumn holiday', making use of local farmers' wagons to visit the surrounding Devonshire countryside; in particular, 'a certain fairy dell'.[282] Though the narrator made banal observations on it being 'a perfect day for a picnic', the pedagogical purpose of the text was not forgotten, even in these more descriptive passages: for instance, 'the botanists of the party' called a halt to the wagons to 'search for the little Sundew

(*Drosera rotundifolia*)': the inclusion of the scientific name ensuring readers could learn something from the description. Such a more story-like presentation, then, also reflected the means by which microscopic specimens could be procured by readers themselves; indeed 'several specimens were uprooted and carefully packed away to plant in moist moss at home'.[283] Upon encountering the fairy ring, the Professor also took 'a few home from where they can be spared from the ring', hoping to 'learn their history' on the morrow, and to show his students 'the wonder-working pixie' who had made it, by examining it under the microscope. This practical setting-up of the subsequent microscopic dissection was combined, however, with Shakespearian allusion, with a quotation from the *Merry Wives of Windsor* on the 'meadow-fairies'; and a comparison between the students themselves and the creators of the fairy rings: 'If we are magicians and work spells under magic glasses, why should not the pixies work spells on the grass?'[284] Just as in the *Fairy-land of Science* forces had been fairies, and gravity a giant, here 'patches of a beautiful tiny mushroom' were revealed as 'our fairies'.[285]

Other nineteenth-century works had traced such precise correspondences between objects of fairy lore and scientific equivalents: for instance, Michael A. Denham's *A Few Fragments of Fairyology, Shewing Its Connection with Natural History* (1859) had also used a Shakespearian quotation when introducing the fairy ring. This work listed many fairy objects,

and gave notes explaining their origins or, he argued, cor-
responding natural objects. For instance, 'fairy butter', number 3 in his list of 60 objects, was glossed with the following
comment: '3. *Tremella meseterica* A substance occasionally
found after rain on rotten wood, or fallen timber; in con-
sistency and colour it is much like genuine *butter*. It is a
yellow gelatinous matter, supposed by the country people
to fall from the clouds. Hence its second popular name, or
star-jelly.'[286]

Back at school, the next section of Buckley's chapter
presented a nature study lesson in action.[287] The Professor
introduced the eager students to the class of organisms
known as *fungi*, using wall diagrams as well as microscopic
specimens, which were placed in the charge of the elder
boys. Fungi, the Professor explained, were 'all mushrooms
and moulds, mildews, smuts, and ferments'. He declared that
'we were not so far wrong when we called them pixies or
imps, for many of them are indeed imps of mischief, which
play sorry pranks in our stores at home and in the fields and
forest abroad'.[288] The fungi, the Professor explained, were to
be found everywhere, and were to be best examined under
the microscope. The accompanying illustrations showed
microscopic perspectives on various stages of mycological
development, and worked with the Professor's lecture to
describe its life-cycle 'in a single night', and its 'secret' underground existence which makes for the fairy ring.[289] This, the
Professor concluded, was 'the true history of the fairy rings,

and now go and look for yourselves under the micro-scopes'.[290] In these ways, Buckley's work built upon her earlier publication that revealed the true wonders in the surrounding natural world, both underfoot and overhead.

In the sky-garden

A range of nineteenth-century children's books, both British and American, brought together fairy tale frameworks with the latest astronomical, microscopical, and pedagogical imagery and techniques. The works often deployed a com-bination of more familiar referents and more imaginative conjecture, a blend of fantastical framing and everyday demonstration which played around with the 'instructive and amusing' genre's narratological and pictorial strategies. For instance, Agnes Giberne's 1884 *Among the Stars* took several tropes from contemporary children's stories and combined them into a chronicle of one boy's astronomical education.[291] The book was explicitly designed as a 'little volume for children' '*much easier*' than her previous intro-ductory work, *Sun, Moon, and Stars* (1879).[292] The 'total eclipse of the lamp', for instance, used similar objects to Buckley's demonstration depicted in Figure 17: a boy and lamp and ball in a domestic miniaturization of the heavenly bodies. Other explanations, however, were more elaborate: summer in the northern hemisphere used a globe that happened to be at hand; or a child literally ran rings around his chair-bound teacher as he physically enacted what an 'orbit was'. Elsewhere the book was at pains to stress there were only certain astronomical phenomena that could be revealed without a telescope: luckily, paying a visit to the observatory of Mr Fritz, a family friend, provided a suitable opportunity to remedy this, with an expertly guided introduction to telescopic sights (see Fig. 18). Furthermore, the book also demonstrated the power of fiction, particularly when combined with more grounded, sensory demonstrations: the ultimate way of learning about the stars was not to bring them down to earth, but instead to fly up to them on wings of imagination. The writing of a story nested within the story of *Among the Stars* itself served to draw attention to this fictive element, the imaginative leap that needed to occur (see Plate 5).

Similarly, American author Lizzie W. Champney's 1877 work *In the Sky-Garden* used fantastically allegorical and imaginative illustrations and accompanying stories to depict what the narrative termed the 'fables of astronomy', drawing heavily on classical myth as well as journeys through the heavens (see Plate 6).[293] The dedication of this book of 'fables' was to 'Professor Maria Mitchell' by the author, her 'pupil and satellite': Maria Mitchell was professor of astronomy at Vassar College in New York, well-known for her telescopic discovery of a comet in 1847.[294] Champney graduated from Vassar in 1869, and named her daughter

IN MR. FRITZ'S OBSERVATORY.

Fig. 18. 'In Mr. Fritz's Observatory', from Agnes Giberne, *Among the Stars: Or, Wonderful Things in the Sky* (1885). Giberne's illustrations moved from the domestic demonstration of astronomical phenomena to the representation of more specialist scientific locations and instrumentation; here, an observatory.

Maria Mitchell Champney after her old teacher; the illustrations to the work were by her artist husband, known as 'Champ': the book was very much a family affair. At one point in the story its young female protagonist, Joy, was—as in *Among the Stars*—taken up to the celestial realm. Though she had travelled to the stars, the narrator noted how 'the strangest thing of all to Joy was, that every thing here should seem so perfectly familiar': just as with the drops of water in Chapter 3, unexpected and novel sights were made comprehensible by comparison with objects already encountered by readers.[295] Here, perhaps riffing on the notion of astronomy as a 'natural history' of the heavens, the new sights Joy encountered were explained using horticultural similes, as in the 'sky-garden' of the book's title. However, the limitations of attempts to learn through the 'present surroundings' were soon revealed, when a lesson went awry:

> Mr. Fairchild began a scientific description and explanation of the phenomenon, just such a one as he would have given to the junior class at the university, but illustrated it from the present surroundings by calling the axle of the front wheel of the carriage the earth, the lantern the sun, and a cake of mud that had adhered to the tire of the wheel, and was revolving rapidly around the axle-earth, the moon.
>
> The only trouble was that the revolutions of the mud moon were so swift that it made Joy nearly dizzy to keep it in view; and her father's demonstration, with its learned allusions to parallax, the sun's photosphere, heliocentric longitude, and the perturbation of the elliptic motion of the moon, was quite as difficult to follow. When she tried to apply it to her own case, and see just how it was that she had been eclipsed, the problem

became still more complicated; and she became hopelessly puzzled...[296]

Joy had been 'hopelessly puzzled' by the 'mud moon'—this demonstration (albeit fictional and played for comic effects) revealed the potential problems with explanations that relied on everyday referents, which could obscure as much as enlighten. Here, however, and as in many of these fairy-tales of science, another option was open to the narrator in teaching Joy about the joys of the universe: fiction, with the incorporation of a classic mode of visiting the heavens, a dream-journey:

> Scarcely had her eyelids closed, when the great, genial Sun reined in his twenty-four-in-hand at her window, and invited her to take a ride. Joy clambered up by his side... She could see a silvery white river gliding placidly through the heavenly fields... it made her think of some views she had seen of Holland...
> 'What river is that?' she asked of the Sun.
> 'It is the Milky Way,' replied her escort.[297]

Travelling alongside the sun in his chariot, Joy saw at first hand the different phenomena of the universe. However, she again compared what she saw to more everyday objects: the Milky Way was in both metaphor and simile a river; more precisely, a Dutch river. Familiarity was put to work, even in the most fantastical imaginings.

These fairy stories of the stars were a significant way of understanding the structure and contents of the universe for nineteenth-century readers. In updated versions of natural magical traditions, students were taught that learning to master technologies such as telescopes, spectroscopes, and cameras was akin to seeing through magic glasses to the farthest edge of the solar system. When combined with fantastical presentations that took to the heavens on wings of imagination, information about suns and planets, satellites and constellations, could best be communicated as part of a pedagogic repertoire that ran from familiar demonstrations with everyday objects, to domestic scientific instruments, to fanciful imaginings. Astronomy was likened to natural history; and set alongside microscopical introductions which taught of the secrets within the local as well as distant realms, and the real history of fairy rings. But mastery of the heavens was just one way in which the confidence of Victorian science and technology was expressed, as new artefacts demonstrated the power of scientific men over the forces of nature, as messages sped around the world, ghostly spectres were conjured on stage, or men could even fly through the skies themselves.

6

Modern Marvels

'Faster than fairies, faster than witches...'
Robert Louis Stevenson, 'From a railway carriage',
A Child's Garden of Verses (1885)[298]

The darkened stage of the Savoy Theatre, 27 November 1882. The star performers of Gilbert and Sullivan's highly anticipated new operetta are making their final adjustments before the curtain rises. Wigs are secured. Cheeks are pinched. Lines are whispered. Shoes are buckled. Throats are cleared. And batteries are switched on. What, you may well ask, were batteries doing on stage at the Victorian Savoy? Concealed within the flowing drapery of the characters' clothing, these oozing, heavy, uncomfortable boxes supplied electricity to the miniature bulbs affixed to headpieces and wands that would amaze the audience later that night: the world's first fairy-lights. Reviews of the Savoy noted the technological accomplishments of the costuming: 'In the last scene a very brilliant and original effect is introduced. The Fairy Queen and her three chief attendants wear each an electric star in their hair. The effect of this brilliant spark of electricity is wonderful.'[299]

These brand-new scientific devices were quite literally to be found lurking beneath the fairy tale surface of the evening's entertainments, but there were many other ways in which the two could be combined in the second half of the nineteenth century. Electric lights were just one of the new technologies that promised to deliver to Victorian audiences what had previously been confined to fictional and magical speculation: conjuring light from the darkness was joined by rapid journeys in seven-league boots, magical mirrors, speaking directly to those in lands afar, even inflating balloons and taking to the skies. This chapter explores the written and performed technological marvels of Victorian Britain, to reveal how they were compared, contrasted, and conflated with the tropes and characters of fairyland. Just as the batteries underpinned the theatrical effects of the stage, so too—authors argued—was access to the fairyland of science like the 'behind-scenes' of a theatre. Scientific wizards, and their inventive devices, allowed a glimpse into the hidden mechanisms of the universe.

Goblins and ghosts

The Savoy's electrical batteries were not the first time the sciences had graced the London stage; not even the first time in fairy tale guise. Over the 1848–9 Christmas and New Year period the Victoria Theatre played host to a scientific pantomime, titled *The Land of Light*, or 'Earth, Air, Fire, Water; Harlequin Gas and the Flight of the Fairies' (see Fig. 19).[300] Following the deregulation of the metropolitan theatres in 1843, the predominant frameworks of the earlier

Fig. 19. Playbill for *The Land of Light*, Monday, 1 January 1849. The playbill details its extraordinary series of scenes, with successive locations taking audiences below ground and to the heavens, to traditional fairy tale locations as well as to local theatres and scientific institutions.

Harlequinade and 'extravaganza' forms were reworked in innovative ways to create the combination of nursery tales, moral instruction, and pointed socio-political commentary that would define the pantomime, and would become an essential festive entertainment for the whole family.[301] Fairies and fairy tales were some of the most popular characters and plots to incorporate into novel pantomime productions: even more so than art and literature the stage was an immediate means of bringing fairyland to life for Victorian audiences.[302] In the pantomimes' picaresque plotlines and elaborate sets and scene-changes these more timeless figures were combined with topical developments to provide commentary on anything from society scandal to new technologies. In 1851, for instance, the Britannia Theatre would show *Harlequin and the Koh-i-noor, or the Princess and the Pearl*, referencing the Indian diamond that had been the prize exhibit at Hyde Park earlier that year: the most famous crystal of the Crystal Palace.[303] By 1865, the Metropolitan Railway (and concerns of steam-driven dangers) would be the star of a crucial scene for the actors in *Old Daddy Long Legs, or the Race for the Golden Apples*: being blown up on a train they reappeared in the 'Realm of Clouds'.[304] Audiences were therefore used to making connections to the most newsworthy objects and projects, and authors met their expectations with ever more specific references, and ever more elaborate stagecraft.

Even in this culture of up-to-date commentary, *The Land of Light* was notable for 'being ambitious of a scientific interest', as a review of the 'Pantomimes of the season' in the *Athenaeum* put it.[305] Indeed, its plot was propelled by the making of a bet on the superiority of science over story: a claim that the 'powers of Science out do [those] of fiction', her 'wonders beat[ing] all fairy ones outright'.[306] Such an emphasis on 'improvement' was characteristic of the oeuvre of Edward L. L. Blanchard, the best-known author of mid-century pantomimes, including *The Land of Light*: his first Drury Lane production (with which he would make his reputation) in 1852–3 brought together 'Antiquity and the Spirit of Improvement' to make the best of past and present.[307] His later plays, like others of the time that were now catering to an explicitly familial market, engaged with current debates over appropriate entertainments for children, and often included a moral message, whether the battle between 'King Nonsense' and the 'Spirit of Common Sense' (which involved a victory over 'Red Tape'), or a tussle between 'Intelligence, aided by Imagination, Discovery and Invention' and 'Ignorance, backed by the spirits of Prejudice and Superstition'.[308] The same discussions that were on-going in middle-class periodicals, or in the prefaces to juvenile printed works, were here transformed into plots for yuletide entertainments.

Following conventions of opening pantomime entertainments in underground or underwater locations, *The Land of*

Light began deep within the coal strata, where exiled fairy favourites and staples of the Christmas stage, including Oberon, Titania, Puck, Robin Good Fellow, and Queen Mab, bemoaned their banishment from the modern world far above them.[309] Echoing some of the concerns we have explored in the Introduction, that the industrial Victorian world had no place for fantastical creatures, they complained of their 'terrible fate, to be placed in this state', 'banished by Modern Improvements'. Their 'only chance', the opening chorus claimed, was 'the downfall of Science'.[310] The scene placed at its centre 'a large slab of Coal', foregrounding the crucial role of the substance in steam-powered contemporary society; stage directions revealed that the subterranean setting was in fact 'the Goblin Coal Mine 5000 miles below the surface of the earth'.[311] The play both joked about the fairies' current situation and also played for laughs, as was traditional: like most pantomimes of the time, much of the script was written in rhyming couplets, and set up elaborate puns.[312] Referencing the coal with which he was surrounded, Oberon, for instance, remarked that it was 'plain we've got the sack'.[313]

It was not long, however, before 'Science' 'found out' the fairies' 'retreat', entering with 'Mysterious sounds' and a 'violent clash' as the 'back part of the scene' broke away to reveal Science and attendant 'navies [sic]'.[314] This appearance in the underground mine was, the ensuing conversation revealed, apparently an accidental discovery, rather than the 'insult' that Oberon claimed: Science and the 'navies' had been 'teaching Brunel' 'how to form a tunnel' when they had happened upon the gathered characters.[315] Tensions that we have explored throughout this book between modern technologies and the consequent disruption of fantastical life were drawn out in a dialogue between Oberon and Science: the fairy claimed that 'our spells aren't half so powerful as yours', and that 'all's up with us when Science once come near us'.[316] The verbal exchange was succeeded by a physical demonstration, one of the many special effects that attracted audiences to these shows: Science ignited 'with her wand' the gas on the surface of the central slab of coal, and 'Gas' then appeared as 'a sprite...with flame upon his head'.[317] As Oberon declared, this was 'Magic indeed'. Science's riposte was cutting: 'Pshaw! Thats naught—Behold' [sic]. With another wave of the wand, the scene changed to a second fantastical setting in 'the Lustrous Land of Light': set directions indicated 'a dazzling radiating Sun at back with glittering wings powder'd with coloured stars and forming a magnificent temple', in which was showcased the latest forms of lighting, including the Bude-light, a type of oil-lamp developed in the late 1830s.[318] It was clear that the show would celebrate the wonders of modernity.

The rambling plot-line and varied scenes of the panto-mime were typical of mid-Victorian productions: as William Thackeray complained, they were 'intricate and difficult to understand'; perhaps viewing the rapidly evolving settings helped compensate for any lack of comprehension over narrative or characterization.[319] *The Land of Light* toured various contemporary venues, both magical and local: traditional fairy tale locations, such as 'The Hut of Walter the Woodman', and a castle exterior; London landmarks such as Drury Lane Theatre itself and the East India Docks; not to mention, like the characters from the Metropolitan Railway accident, ascending to 'the realms of Cloud land in the territory of twilight'.[320] The surviving manuscript gives some hints as to how the engineering feats that accompanied the transitions between these scenes were managed, as when 'Scene 5th' is marked with 'a Front Cloud to give time for Balloon to get half under stage—purple balloon to work on Flats'.[321] Other surviving accounts also reveal the technological underpinnings to these changes of scene and evocation of particular locations, as well as how they were worked into the performances: one waterfall of real water, for instance, took eight mechanical changes and fifteen minutes to be put into effect; elsewhere, gas jets and globes of cut-glass were deployed to create a spectacular rendering of 'the Submarine Grotto of Phosphorescent Light'.[322]

'Land of Light''s whistle-stop tour culminated with the scientific sights of the metropolis, in what was termed, in a moniker bringing together the pantomime's main themes, 'The Hall of Magic Science'. The pantomime audience would have been familiar with venues such as the Adelaide Gallery or Royal Polytechnic Institution: mapping onto the leisured geography of London, from fashionable West End haunts to cheaper working-class attractions, these like the pantomime were part of the wide range of scientific entertainments on offer in the metropolis.[323] *The Land of Light* was advertised alongside rival venues' entertainments at Drury Lane, Sadler's Wells, or the Strand, and in the theatrical columns of weekly newspaper *The Era*, with boxes to be had for 1s., 6d. for the pit, and 3d. for the gallery; but on the very next column, lying just to the right of the Victoria's announcement, was an advertisement for 'Christmas Holidays' at the Royal Polytechnic Institution on Regent Street: for 1s., or half price for schools, patrons could hear 'A LECTURE on the POPULAR SUBJECT of the ELECTRIC LIGHT, or BRILLIANT EXPERIMENTS, DISSOLVING VIEWS, microscopes, and, its highlight, a working model of a DIVER and DIVING-BELL'.[324] Demonstrating such novel technologies as the oxy-hydrogen microscope or enormous induction coils, these shows also celebrated these new technologies, cast their presentations along magical lines, and supported the rhetoric of the fairy tales of science with their cognate concern for the revelation through explanation of scientific wonder and power.[325]

No scientific wonder of the stage brought out these concerns better than Pepper's Ghost, developed by John Henry Pepper, who had been appointed to the Royal Polytechnic Institution in 1848, the year *The Land of Light* opened, as lecturer and analytical chemist.[326] The relationships between the venues for scientific demonstration and theatrical show were personal as well as topical and thematic: Pepper also worked with the theatres themselves on the stagecraft of their shows, for example providing the 'scientific arrangements' and scenery for The Diamond Maker in 1865.[327] But Pepper's Ghost was to be his enduring legacy. First demonstrated at the Polytechnic in 1862 to accompany a performance of Edward Bulwer-Lytton's A Strange Story, the illusion used an enormous sheet of plate glass, placed at a 45° angle to the stage, and a bright light: this apparatus conjured a reflection of an actor standing in the orchestra pit, which appeared in phantom form on the stage above.[328] The spectre could then be interacted with by the rest of the troupe, to dramatic effect: its use at the Britannia Theatre in 1863, for instance, in displaying an altercation between a baronet and a murdered widow, rendered the audience 'spell-bound', 'unable to explain the cause of so extraordinary an appearance', when the figure on stage was seemingly run through with a sword.[329] However, as scholars have shown, Pepper used his ghost as a means of educating astonished audiences about the optical principles underlying his illusion, explaining how the visions had been created. Eminent visitors to his Institution, such as the Prince of Wales, could even be taken behind the scenes to witness Pepper's wizardly mechanism that produced the ghost itself, revealing the rational principles that could both mimic and arguably surpass the magic tricks of old.[330] From goblin coal mines to ghostly contrivances, these scientific shows were an important space in which debates about who should be the authority on the natural world were conducted.

Grander tours

A favoured fairy tale motif has been the seductive power and endless possibilities of supernatural travel: magical artefacts such as seven-league boots that made journeys pass in the twinkling of an eye, as seamlessly as one of the scene-changes in a pantomime. Thanks to developments in steam-driven transportation, by the mid-Victorian period it seemed that one could travel the globe even faster than the newly electrified Puck could girdle it (see Fig. 20). John Cargill Brough made direct comparison between Puck and the telegraph in his *Fairy-Tales of Science*, lauding the 'Amber Spirit' of electricity which, 'in spelling a word of one syllable...has to perform a series of journeys which together far exceed the length of Puck's famous girdle'.[331] Man, Brough argued, was now capable of controlling what had hitherto been a phenomenon more suited, from ancient marketplaces to eighteenth-century salons, to the demonstration of the mysterious powers of nature, as his identification of this as the 'Amber Spirit' evoked.[332] By the

Fig. 20. 'Puck girdles the world', an innovative illustration by Charles H. Bennett to John Cargill Brough's *The Fairy-Tales of Science* (1859), representing the electric telegraph as Puck. The invisible force of electricity was often depicted as a sprite or imp: the Shakespearian allusion makes Puck a particularly appropriate choice here.

mid-nineteenth century, it was clear that electricity could be put to work 'as a courier, a vocation for which he is eminently suited', part of imperial projects that spanned the globe.[333] Published in 1859, Brough's book reported on the very latest in technological achievement, including the laying of the transatlantic telegraph cable, which had occurred with brief success the year before:

> Europe was covered with a network of wires, and so was America—to unite these two great systems of communication would be a feat unparalleled in the annals of Science....This wondrous feat has at last been accomplished, and the two great Continents are now connected by a cable which lies at the bottom of the Atlantic. At Man's bidding the Amber Spirit speeds along this tremendous cable, and having registered a single letter at its further end, finds his way back to the battery through the pathless deep...[334]

As Brough's discussion revealed, modern technologies of communication and transportation underpinned this imperial age, in a 'network of wires' and shipping routes that tied together disparate colonial territories. 'We may search through our old fairy tales and romances in vain to find a spirit capable of performing such miracles as these', Brough wrote.[335] Harnessing that imperial mindset and connecting it to both modern technologies and fairy tale artefacts, writers could go far beyond the simulated grand tours of the early century to take their readers on even grander tours.

In the *Children's Fairy Geography*, written in 1879 by the Reverend Edward Forbes Winslow, and republished in a second edition in 1885, electrical technologies complemented and also replaced magical fairy tale devices for keeping warm, making light, and, especially, for travelling a great distance in a short amount of time.[336] Ostensibly a guide to the scenery and inhabitants of the various European nations, *Fairy Geography* was also an introductory textbook built around a travel narrative, and suffused with humorous depictions of modern marvels. In the preface its author was at pains to justify the incorporation of the supernatural as well as the entertaining aspects of his 'merry trip' (the book's subtitle), which—he worried— might not sit obviously and easily within an educational geographical primer. He protested that some of his readers 'may take exception to the fairy element, others may say that the jokes are weak, and poor, and very very old (a true bill I most humbly confess)'.[337] But it would be impossible to please everyone with the precise balance of facts and fancy: 'some may complain of the number of dry statistics, others may blame me for not having more, one kind friend may applaud, and another equally kind, may condemn.' Therefore, he had chosen to combine weak jokes and dry statistics in his own way: such a combination, he thought, befitted a newly technological age, one in which even the toys of the nursery had been taken over by 'elaboration', converted into 'walking and growling bears, talking and swimming dolls, and [even] mechanical puppies'.[338] This tale, then, would add to the new kinds of 'school-books' which were full of 'life, animation and interest', by taking its protagonists on a fantastical voyage.[339]

Although recent scholarship has identified how common the combination of fact and fantasy was in elementary geography books of the late nineteenth and early twentieth centuries, since fairy tales' 'fantastic conventions made them a useful means of travelling, in imagination, to distant and exotic locations', contemporary reviewers were not convinced.[340] As Winslow had anticipated, one critic passed a damning verdict in the *Spectator* on the entire pedagogic project of combining instruction and amusement:

> The pictures are certainly very numerous and good, but the letter-press fails in being either attractive reading on the one hand, or systematic and exhaustive geography on the other. It attempts a compromise, and fails, as compromises of this sort always do.[341]

Rather than these combinations of fairy tales and science being a beneficial combination that was more than the sum of their parts—emphasizing the right kind of imagination; providing better monsters and more miraculous transformations—for this reviewer the writer had attempted to have his cake and eat it; and it was a poor cake to begin with. The reviewer attacked Winslow on both counts: not only was his relevant content meagre and 'sketchily' introduced, but the 'fairy part of it, too', was 'so transparent and thin a make-believe, as to provoke rather than amuse an intelligent child'.[342] The framing device chosen by Winslow, inviting his childish protagonists to embark on 'a new way of travelling, which will make you open your eyes', was a well-worn metaphor, and echoed Arabella Buckley's assertion that the way to the *Fairy-land of Science* was through opening one's eyes to the wonders of the surrounding environment. With many illustrations as well as a similar visual rhetoric to Buckley in his prose, Winslow sought through his work to vicariously offer first-hand sight of the sites of Europe to his readers. The reviewer, however, complained that most of the sights that Winslow claimed to show the child were not actually possible at all. When, he wrote, the tutor asks the children, 'Do you see those big barges by the side of the river?', the child 'of course' cannot see them.[343] Purportedly realistic sights were, it turned out, just as illusory as fantastical creations.

More charitably, the text—like others of the time which relied on existing knowledge and everyday objects—was at pains to connect its literary representation to possible embodied practices and observations, by referring back to sights that the reading child had potentially already experienced. The story began with the teacherly narrator and his young charge Ethel embarking upon 'make-believe travels' around England, with the aid of numerous illustrations and imagined trap and train journeys around the country that would have resembled the kinds of excursions a middle-class family like Ethel's embarked on in the 1870s and 1880s.[344] The narrator directed Ethel's sight, pointing out particularly interesting places, objects, and occupations; a map was used to introduce the main manufactures of each of England's forty counties, from Barnsley linen to Cornish copper mines.[345] Indeed, in the preface, Winslow had recommended the use of maps in conjunction with his book, urging that 'this geography will be made more interesting and instructive if maps and plans of the countries and cities mentioned are freely used'.[346] This incorporation of other teaching aids, as well as the book's numerous engravings of buildings, vistas, and local inhabitants, attenuated the reviewer's concern that the children did not actually see such objects themselves. The interplay between everyday and imagined sights was present throughout the book, with the most modern technologies being discussed as fairy tale creations: like the telegraph, the railway train was held up as a magical means of transportation: 'Look at all the lights in the high houses on the further side; the red, white, and green lamps of the trains as they puff backwards and forwards. It seems almost like fairy-land.'[347]

In the subsequent chapters of the book, however, the characters in the story did not merely 'make-believe' they were travelling by train, instead voyaging on the very latest 'little invention, known as yet only to you and me': the 'wishing-carpet' (see Plate 7).[348] Just like magic carpets of the *Arabian Nights*, this kind of supernatural device was very probably familiar to the young readers of the work; moreover, it once more recast a quotidian object in fantastical guise, as the rug beneath the children's feet took to the skies, whizzing from nation to nation and providing direct experiences of various countries (intensely coloured by contemporary concerns). Comparisons continued to be made to more usual ways of learning geography that would have been possible for young readers: for example, when the trio 'cause our wishing-carpet to float so exactly over Scotland, that we can obtain what is called a bird's-eye view of it, so that the whole country shall be spread out under us like a map'.[349] Standing above their spread-out map, reading children could have this perspective themselves.

In his account of European travel and bird's-eye views of the countryside, Winslow drew a contrast but also an inevitable comparison with another means of flying above and about the clouds: ballooning.[350] Whereas the railway train promised power and steam, noise and speed, the balloon was presented as a tranquil alternative: the pastoral counterpoint to the industrial machine.[351] Such peacefulness was—of course—illusory: almost every account of balloon launches from the late eighteenth and early nineteenth centuries had ended in disaster and despair, with passengers tumbled to the ground, carried out to sea, or, worst, sent up in flames. Winslow was not unaware of the comparison inherent in his choice of airy vehicle, and explicitly emphasized the dangers of ballooning versus the 'perfectly safe' carpet, although the reason the carpet was 'perfectly safe' was that it was perfectly imaginary; or, as Winslow put it, 'all make-believe'.[352] But the early history of ballooning was indispensable for the kinds of sights to which the narrator of the *Fairy Geography* guided his readers' eyes: most clearly in this adoption of the 'bird's-eye view' perspective, and in the depiction of the book's characters hovering above a pastoral scene: the view from the magic carpet is exactly like the view from a balloon.[353] The novel perspective on the earth provided by a balloon ascent has been much discussed, and the use of images of wings, butterflies, and fairies as analogues for the sensation of flight has also been researched, not least with respect to pictures of flying fairies from the period.[354]

The account of the children's tour of Europe reads very like those detailing early balloon flights: in particular, the overnight ascent of renowned balloonist Charles Green in 1836. Launched from London's Vauxhall Gardens at 1.30 p.m. on 1 November of that year, his 'Nassau Balloon' was subsequently to be christened after the destination the party reached at 7.30 a.m. on 2 November, in north-west Germany: eighteen hours of continental flight.[355] Like the characters in Winslow's book, the inhabitants of the Nassau Balloon's basket travelled from their home, through the clouds, to a nearby geographic location. Newspaper reports gleefully detailed the preposterous amount of provisions with which the three men stocked their balloon, from meat and game to port and champagne, ropes, maps, and telescopes, not to mention Green's coffee-maker, which combined chemicals to heat its contents rather than using an open flame.[356] As well as telling of the marvellous prospects from the balloon, and retracing its path across Europe, the balloonists also included mention in their accounts of a curious trick they had played as night fell on the darkened towns they passed. Lowering their lamp from the basket of the balloon, they amplified their voices with a speaking trumpet to make pronouncements in French and German to the earthbound inhabitants. Not content with such a quasi-supernatural visitation, they then proceeded to pour some of the sand they carried for ballast onto the figures below, claiming to have left them 'lost in astonishment' and fear.[357] Just as the appearance of Pepper's ghost had baffled and awed its first audiences, but was explicable upon further investigation and upon scientific principles, here again were the uneducated onlookers held up as believing that modern marvels were in fact magical apparitions. Astonishment was the marker of a neophyte reaction.

In Winslow's tale, the wishing-carpet and its ballooning analogues was just the first of an array of new-fangled devices brought into the narrative to help the children and their teacher on their journey around Europe, from 'Yachting Telephone' to 'portable Boyton's dresses', referring to the inflatable rubber suit worn by Paul Boyton for his headline-grabbing exploits in the water around this time.[358] The narrator lampooned the tendency for pseudo-classical names for many new technologies of the nineteenth century (including telephone, telegraph, and phonograph), with his choice of name for one of the novel artefacts. Introduced while the travellers are in Holland, one of Winslow's self-confessedly 'weak jokes' is made: that 'the new *Micromegalopanthaumastophonograph*' is 'double Dutch' rather than what it should be in 'plain English', namely 'a sound-recorder'.[359] Handily available to rescue his protagonists from the ill-effects of international travel, the inventions brought into the narrative also conjured something of the atmosphere of global exploration that underpinned the geographic subject of the book, not to mention the numerous patent contraptions on sale in the metropolis to allegedly ease the life of the provincial Victorian sent out across the empire. For instance, towards the end of the book when Jack, Ethel, and the guide traversed 'The Arctic Seas' (and inscribed their names on the North Pole), the adult produced 'just the thing to make us all comfortable' in the icy wastes:

> Do you see this round thing, something like a china cannon-ball? It is called *The Chilly-warmer*. I take two small bottles of a chemical preparation that I carry in my pocket, and pour a few drops of each into these two tiny holes you see at the side. These liquids meet together inside the apparatus, and immediately, by their action one upon another, they make the Chilly-warmer so warm, that even at a good distance off you can feel it like a cosy fire. Once set in action, the warmth will continue for hours. But I have not yet done with the wonderful properties of this new invention. It is called the *Chilly-warmer* not merely because it warms the chilly, but because it also acts as a chiller,

or refrigerator. By putting two other liquids inside, I make a freezing mixture, which, of course, makes the machine as cold as ice. And so, if we were travelling in very hot countries, and we wished to have our room cool and comfortable, all that we should have to do would be to freeze our Chilly-warmer. When not used either for warmth or for cold, I can unscrew the two halves, and then we have a couple of basins for making tea, or drinking out of.[360]

Not only capable of keeping travellers warm or chilled, the miraculous device could also be used to furnish its owners with the fuel on which the empire ran: tea. (Discussion of this device also recalls the patent coffee-maker taken by Green and his colleagues on their ballooning adventure, which sadly fell overboard at one point in the lengthy flight.[361]) Such a colonial subtext can be traced throughout the work, from its sneering depiction of Ireland to its assumed superiority of the English middle-class travellers and their technologies over the lands they visited. At one point a variant of the 'electric light' 'in the shape of a candle' which 'the moment it is blown out re-lights itself' is explicitly said to be something that would 'astonish the natives'.[362] Again, comparison can be made with the exploits of the Nassau three, in their attempts to astonish and terrify the Belgian population by judicious manipulation of a safety-lamp. A suitable scientific education, it was clear, was the only means by which one could appreciate such feats as technological triumphs.

As well as the British Empire, the wider Anglophone world was also referenced in the work: just as the transatlantic cable had tied together Britain and America in 1858, so too was the United States the specified source of several of these novel devices, including the 'Micromegalopanthaumastophonograph' (perhaps a reference to Thomas Edison's very recent invention of the phonograph), and the 'Electric Boots, or Speedaways' (see Plate 8). The 'Speedaways' were a literal updating of Jack the Giant-Killer's seven-league boots, unveiled in Winslow's *Fairy Geography* by a character who shared his name:

> Jack, what are you rummaging about in the box for? 'Only three pairs of boots!' You need not speak in such an injured tone. But let me just observe that those *are* boots. Look at the wires running down the sides, and the mechanism in the heel, now that I have unscrewed this one for your inspection. These are Electric boots, or Speedaways. When we put them on we can leave the seven-leagued boots of the old fairy tale far behind us. When we get into Switzerland we will have an excursion in them, and I know they will astonish you. I see these that have been sent to us have been marked '15;' that is to say, they are only guaranteed to walk fifteen miles an hour.[363]

With a swish brand-name, these new boots were overtly modern, and—as Winslow's narrator claimed—left 'the old fairy tale far behind'. The children were also identified with the 'natives' as another audience who could be 'astonished' by the products of modernity. For something that was on one level a mere narrative device to enable the swift passage of its characters from one location to the next, these new electric boots were very precisely described, even down to the speed with which they enabled one to travel. Just as elsewhere Winslow relied on the referencing of familiar sights or the incorporation of maps and carpets at hand in the home, here the possible reality of the story's events was again teased out. The 'wires running down the sides' and 'the mechanism in the heel' were emphasized, demonstrating that the narrator knew the scientific and technological principles at work in the boots. His control over the technology was again demonstrated in the following section:

> By twisting this little regulator, we can put on additional electric power, and get a current so strong that we stride through the air without touching the ground at all. Never mind the tingling, Jack; you will soon get accustomed to the feeling. It only arises from the currents of electricity that are at work under the sole, and in the heel. Now, are you ready? Off we go.[364]

Electric devices, it was clear, could give one the power to travel at unimaginable speeds, and in unfamiliar ways; but electricity was a powerful force to tame, and one whose side-effects could be much worse than mere 'tingling', as the hero of a later children's book was to discover.

Wizards and demons

The Master Key of 1901 was avowedly 'a fairy tale founded upon the mysteries of electricity and the optimism of its devotees'.[365] Ushering in the new century, the tale was, its author L. Frank Baum declared, written for 'children of this generation', who, living more than thirty years after the publication and setting of Winslow's Fairy Geography, would have been much more familiar with electrical objects and the future's electrifying potential. Indeed, Baum predicted that the succeeding generation thirty years hence would live in yet another world, in which his 'story may not seem... like a fairy tale at all'. The 'impossibilities of yesterday' would 'become the accepted facts of to-day'.[366] The story claimed for twentieth-century children hitherto unimaginable control over the forces of nature, its narrative detailing how a young boy called Rob inadvertently conjured the Demon of electricity, who gave him privileged access to electrical devices (see Plate 9). Baum was, children's literature scholars have shown, 'obsessed' with the potential and the perils of electricity, dealing with it in several of his works, including the later Oz books for which he would be most remembered.[367] In the Master Key, Baum explored most closely the possible dark side to the Faustian bargain struck by those seeking greater control over the forces of nature, seeking to—as Brough had detailed in the Fairy-Tales *of Science*—put them to work; or, as Winslow had demonstrated in *Fairy Geography*, inventing ever more elaborate devices with which they could be manipulated.

The book opened with a celebration of the young child fed a diet of the kinds of books we have explored throughout *Science in Wonderland*: a boy at the end of the nineteenth century who was fascinated by electricity, and had a demonstrable 'fancy' that 'his clear-headed father considered... to be instructive as well as amusing', and something to be 'heartily encouraged'.[368] Rob, therefore, growing up with a supportive parent and in a suitably middle-class home, 'never lacked batteries, motors, or supplies of any sort that his experiments might require'.[369] He set up a telephone network to each room of the house, with electric bells on their doors and alarms on their windows; this bothered his mother and sisters, but his father thought it was the sign of budding genius:

> 'Our boy may become a great inventor and astonish the world with his wonderful creations.'
> 'And in the meantime,' said the mother, despairingly, 'we shall all be electrocuted, or the house burned down by crossed wires, or we shall be blown into eternity by an explosion of chemicals!'[370]

The maternal warning-note should, perhaps, have been heeded. Messing about one day in his home laboratory, Rob was startled by a 'splendid apparition', who 'bowed and said in a low, clear voice: "I am here." "I know that,"

answered the boy, trembling, "but WHY are you here?"[371]
The 'Demon', as the apparition became known, was a genie-
like spirit for the child genius, a fairy tale reference made by
the story itself, as Rob commented:

> 'But it seems you're more like a genius, for you answer the
> summons of the Master Key of Electricity in the same way
> Aladdin's genius answered the rubbing of the lamp.'
> 'To be sure. A demon is also a genius; and a genius is a
> demon,' said the Being. 'What matters a name? I am here to do
> your bidding.'[372]

Once more the scientific reworking of a traditional figure
was deemed superior, whilst simultaneously being evoked
as a useful narrative form and plot device: the Demon went
on to tell Rob that due to his Aladdin-like role he was also
able 'to demand from [him] three gifts each week for three
successive weeks'. The only caveat was that the gifts must be
'within the scope of electricity'. The pun on the shared
etymology of 'genius' and 'genie' allowed for a digression
on two American figures given such a moniker: both Edison
and Tesla were discussed by the text, in humorous throw-
away references that indicated an expectation that boyish
readers were familiar with their names and most famous
accomplishments. For instance, Edison had 'invented no end
of wonderful electrical things' (dismissed by the Demon as
'trifling'), and Nikola Tesla was apparently 'in communication with the people in Mars'.[373] The Demon confessed to
having stalked Edison's own laboratory, willing him to

touch the 'Master Key' that would summon him to his side, but regretfully concluded that the inventor would never, unlike this precocious child, have the requisite ability.

In total, the Demon gifted Rob with six different electrical devices, and the succeeding chapters of the book narrated his experiences with each, from nutritive tablets which contained a day's electrical nourishment, to a Taser-like tubular weapon, to a 'little machine' that sat like a wrist-watch and enabled Rob to travel up in the air and around the world, just as Winslow's wishing-carpet and electric boots, or the Nassau balloon. The second set of three gifts included a defensive shield, a means of automatically detecting people's character, and a screen with which one could make a record of the age. Despite the novelty of these gifts—foreshadowing as they do many later twentieth-century inventions—the narrative of the book was more old-fashioned, reliant on timeworn plots such as cannibal islands, pirate ships, Cockney policemen, Middle Eastern caravans, or shipwrecked sailors, resembling the stories found in many turn of the century boys' periodicals and book series. At moments, the wider scientific context was incorporated into these quixotic adventures: for instance, one chapter detailed Rob's encounter with a French man of science, who had spotted Rob through his telescope, and could not conceive of how he could be flying through the air.[374] Towards the end of the novel, the story took a darker turn, to comment on the lengths some people would go to to procure patents for these new devices: back in the United States, an old man reading an article on 'The Progress of Modern Science' tries to kill Rob and steal his electrical devices, after offering Rob millions of dollars to buy them from him.[375] Ultimately this attempt on his life confirmed Rob in his decision not to prolong his relationship with the Demon, and to refuse any further benefactions. Chastised by the Demon for not making better use of his preternatural 'gifts', Rob blamed the 'infernal' devices for his 'troubles', and ended up feeling isolated by the power and intrusiveness of the objects. The book concluded with the rejection of the Demon's gifts, Rob preferring to remain with the current devices of the age:

> 'Some people might think I was a fool to give up those electrical inventions,' he reflected; 'but I'm one of those persons who know when they've had enough. It strikes me the fool is the fellow who can't learn a lesson. I've learned mine, all right. It's no fun being a century ahead of the times!'[376]

There was no short cut, the story showed, to the acquisition of these kinds of devices: they showed up the unstable status of many current electrical technologies, and could be incorporated into philosophical discussions about anything from data protection to espionage, as well as the disquieting status of the maverick individual inventor when given the power to meddle in nature.

Behind the scenes at the theatre

At the dawn of the twentieth century, just before he wrote *The Master Key*, Baum had introduced his young readers to *The Wonderful Wizard of Oz*.[377] Like many of the introductory guides I have discussed in this book, the tale's premise hinged on revealing the hidden work, inventive powers, modern technologies, and scientific personnel that lay behind the apparent wonders of the world. Throughout her adventure down the Yellow Brick Road, Dorothy had been in awe of this quasi-mythical figure, and was promised that in a land of flying monkeys and cheerful munchkins, he alone had the power to return her to Kansas. But upon reaching the Emerald City she was to be disappointed: as she failed to be convinced by the staged scene before her, the truth behind the eponymous 'Wizard' was laid bare.[378] The 'great and powerful' ruler, who claimed to be 'everywhere', 'invisible' 'to the eyes of common mortals', was, it turned out, merely an ordinary man with a few tricks up his sleeve: he might as well have had the Savoy's electric batteries hidden in his clothing, or been one of Professor Pepper's ghosts.[379] Ingenuity and technological superiority had enabled the Wizard to attain his status, highlighting the reputation that such inventive figures had more generally in non-fictional turn of the century America, individuals who were often figured as 'Wizards' themselves.[380] It was no coincidence, then, that the Wizard both arrived and left Oz in a balloon: a symbol of modernity, invention, and adventurous individuals; of what one early French observer described as 'myths...come to life in the marvels of physics'.[381] In these ways, new technologies, including that of the balloon, but also of electrical telegraphs, gas-lighting, and more, prompted celebrations of the power of the sciences and of those wizardly men of science, but also concerns over their potential misuse. Just as in fairy tales and the Land of Oz, powerful objects could be used to conjure spirits both good and ill: demons as well as genies. The discussion of such objects revealed new possibilities to control invisible forces, or transcend previous limits.

The celebration of the power of the Wizard was at odds, however, with the disappointment felt on the mundane revelation of the balding, middle-aged Midwestern man who was just as ordinary as Dorothy herself.[382] Such a disappointment can also be felt in earlier discussions of 'The Fairy Land of Science', particularly an 1862 article in the *Cornhill Magazine* by James Hinton. Perhaps audience members who were taught to see through Pepper's ghost were also as disenchanted as they were educated. Hinton noted at the outset the common trope that 'the achievements of modern industry' embodied as well as 'surpass[ed]' 'the imaginations of the youthful world', comparing— amongst others—the 'shoes of swiftness' with the 'railway train, to the manifest advantage of the latter', or 'Aladdin's ring' with 'the electric telegraph'.[383] Modern marvels did supersede but also curiously reincarnated their fairy tale forebears. Hinton went on to complicate this picture even further, decrying the 'clouds' that gathered above this scene of modern superiority:

> A pretty fairy-land our science has brought us to. It is like the 'behind scenes' of a theatre. There are all the fine things we admired so innocently at a distant view; we can't deny that we have got them; 'but oh, how different!' The dazzle, the sparkle, the romantic glory, where are they?... Is all the world a stage?[384]

Just as Dorothy had been impressed by the 'dazzle' and 'sparkle' of the Wizard's Emerald City at first glance, but became disillusioned upon a closer inspection, Hinton spoke of the sciences as revealing the innermost workings of nature: the fundamental principles upon which the universe operated, from forces to particles to chemical transformations. As one academic puts it, 'what we thought was magical, ineffable, or illusion is in fact worked by a stage machinery of levers and props, forces and physical laws'.[385] Just like Pepper's Ghost, or the scientific pantomime of *The Land of Light*, scientific ability now permitted the production of illusions and spectacular demonstrations, in particular the re-creation of supposedly supernatural occurrences. However, the educated witness would instead experience what Hinton described as a 'doubled' kind of vision, seeing but also seeing through what was on show.[386] Hinton—and

many other nineteenth-century figures—claimed that comprehending the mechanisms and scientific phenomena
underpinning the natural world led not to a dismantling of
awe but to a greater appreciation of the 'unfathomable
reality' beyond, whether metaphysical or divine.[387] In Oz,
too, magic as well as inventiveness existed as an alternative
force in the universe: in the end Dorothy had no need of the
Wizard's balloon to return home, having had the power in
her silver shoes all along. The reality had been unmasked,
but the wonder remained.

CONCLUSION

Stranger than Fiction

'...logic is often so much stranger than imagination.'
A. M. Low, *Science in Wonderland* (1935)[388]

I f Tennyson's appeal to the 'fairy-tales of science' has been
one poetic motif that has run throughout this book, then
Byron's claim that truth is 'stranger than fiction' has also
rung clear. From its first appearance in *Don Juan* the phrase
was one way in which audiences could conceive of the
novel occurrences of the quotidian world, and begin to
differentiate out story from reality, truth from fiction.[389]
Of course, the claim made by many of the works we have
encountered throughout *Science in Wonderland* was that the
best kinds of writing did not require the need for such
differentiation: the fairy tales of science were a new literary
genre in which factual accuracy did not have to be sacrificed
for enthralling narrative or engaging characters. In fact,

their veracity buttressed their superior status as providing not just replacements for traditional tales and creatures banished by modernity, but improvements upon legends of dragons, genies, and pixies, magic lamps, mirrors, and carpets. However, the very reliance on and comparison to fictional entities demonstrated that these novel scientific productions—be they animal, vegetable, or mineral; phenomenon, substance, or property—themselves had a rather precarious existence. Out of sight in time and space, or only rendered visible through specialist instruments, practices, or rational processes, it was in fact the closeness of invisible forces and monstrous beings to their fairy tale forebears, rather than their difference, that was striking, and that needed to be acknowledged and superseded.

Therefore, these fairy tales of science also have significance beyond their use in and as elementary instructive works: they speak to and can interrogate wider changes in scientific practice, representation, and authority in Victorian Britain. Assertions of the truth of these novel explanations for natural phenomena did not just shore up claims for the necessity of teaching 'nothing but facts' to modern children, but also reveal an anxiety over whether novel theories, discoveries, or instruments did indeed provide access to true knowledge about nature: was a fact something made-up itself? By these explicit comparisons to the fantastical nature of facts, children's authors used this mode of literary representation to participate in wider debates over and quest for scientific authority in the period: a quest that, by the early twentieth century, would largely have been achieved. These types of writing and thinking, most overtly demonstrated in these works for children but present as framing devices and analogies, passing reference and playful comparison, in expert writings and lectures, played a crucial role in complicating the relationships between the factual and fantastical, not only in scientific communication but also in scientific practice. They provided a crucial foundation, then, for the challenges to this confident edifice of scientific truth erected by century's end which were to come soon after. When, with the development of relativity theory, quantum mechanics, and unparalleled power over the natural world, logic really did seem stranger than imagination.

Nothing but the truth

These scientific stories were, their authors argued, superior to traditional tales, providing better fodder for the fancy: for Brough, for instance, they could 'transcend the wildest dreams of the old poets'.[390] For Buckley, her descriptions of fairy forces were 'ten thousand times more wonderful, more magical, and more beautiful in their work, than those of the old fairy tales'.[391] These works, their authors argued, were superior because of their truthfulness:

I thoroughly believe myself, and hope to prove to you, that science is full of beautiful pictures, of real poetry, and of wonder-working fairies; and what is more, I promise you they shall be true fairies, whom you will love just as much when you are old and greyheaded as when you are young; for you will be able to call them up wherever you wander by land or sea, through meadow or through wood, through water or through air.[392]

The 'true fairies' and 'real poetry' of scientific investigation, Buckley stressed, were no childish fancy but a lifelong companion, enhancing every subsequent interaction with the natural world. This truthfulness was what would set apart the fairy tales of science from the myriad other new moral, lyrical, local, natural, or otherwise socializing stories written or translated in Victorian Britain (see Fig. 21). Charles Kingsley made this point towards the end of his *Water-Babies*, in a lyrical passage that hymned the 'tale of all tales' that was the story of nature, playfully suggesting to his readers that they are 'not to believe a word of it, even if it is true'.[393]

But—as ever—much more was at stake and in play beneath this surface rhetoric. As other authors acknowledged, it was not at all clear that the scientific writers' confidence in having reached, and hence being able to communicate, the truth about nature was justified. For Hinton, for instance, an 'unfathomable reality' lay beyond even that which had been revealed by the expert work of scientific men; the access to the behind-scenes of the theatre, in his favoured metaphor, still left inaccessible mysteries.[394]

Fig. 21. 'The Wonderful Lamp', from John Cargill Brough, *The Fairy-Tales of Science* (1859). This illustration of the 'lamp of science' complemented Brough's discussion of the enlightened power of modern scientific enquiry, which could be put to work for the benefit of mankind.

Indeed, the word 'truth'—especially in relation to the sciences—was itself the subject of much speculation throughout the nineteenth century, as part of discussions over objectivity and truth-to-nature; who had privileged access to the truth (men of science or of God? those with specialist training and equipment, or instincts and insight?); or whether uncovering the truth about the history of life on the planet, for instance, was even possible. There was a similar anxiety at the heart of many of these debates: that facts, too, were fictions, things made up by human beings; that 'invention' was not only a fictive, imaginative enterprise, but also a technological endeavour.[395] As scientific disciplines built reputations, institutions, and personnel, they often made comparisons to and claimed the territory of other ways of explaining the world: as we have seen in the case of fairy tales, this was often due as much to their similarity as to their differences.

Science in wonderland

Mine is not the first book to have been titled *Science in Wonderland*. In 1935, a book of the same name was published by engineer and inventor 'Professor' Archibald M. Low, which took its heroes John and Betty on a whirlwind adventure through modern transportation, the streets, the body, the atmosphere, and even the solar system.[396] Crashing their bikes into an omnibus, the 'twentieth-century twins' were amazed to have fallen not through the looking-glass, but rather through the bus radiator, during which process they were converted into drops of water. As in the introductory works we have analysed throughout these chapters, Low's *Science in Wonderland* deployed fairy tale tropes to communicate the latest in scientific knowledge to his young readers. In phrases reminiscent of Buckley and many other writers, he emphasized the wonders of the everyday world: 'It is because I am sure that children ought to like fairy stories that I have written one in which there are no fairies; in which the characters are so true to life that you can see billions of them every day.'[397] For Low, too, this was a domestic tale of everyday wonder: 'we do not have to travel to where the rainbow ends to find fairyland. We live in it every day of our lives.' However, Low went on to give his scientific fairy tale a different gloss to the Victorian examples: he was going to be accurate, rather than true. 'Pilate's question', what is truth?, 'regarding truth remains unanswered', he wrote, but he had 'tried in this book to be accurate', not describing 'a molecule as doing anything that a molecule could not do according to our knowledge'.[398] Though he would give his protagonists *Alice*-like adventures and enlivened interlocutors, Low was clear that his overt use of literary devices was a metaphor for scientific understanding: 'I make no excuse for allowing molecules and raindrops to take part in conversations when we "know" that they cannot talk. Do we really know that they cannot talk? It would be much more correct to say that we cannot understand them.'[399]

By the early twentieth century, the tropes of the fairy tales of science had been cemented, and works relied upon the success of the previous century in making scientific disciplines and personnel into authorities on what once happened and what would happen in the natural world and in the universe.[400] There had been a move away from the Victorian obsession with fairies to, for instance, the more up-to-date emphasis on 'molecules', but the interest in presenting scientific knowledge in such imaginative ways remained. Indeed, it was increasingly the only way in which the latest scientific knowledge could be written about. Low emphasized the difference between the nineteenth and twentieth centuries with an effective comparison (and sly dig at the clichéd conversations that adults often have with children) between his twin protagonists and their parents:

> people always said that John was the image of his father. How could he be? His father was born in the last century, and when his parents were sixteen a fast crossing of the Atlantic meant a journey of two weeks ... in the way they thought, they were as different as the proverbial chalk and cheese.[401]

Just as Baum suspected *The Master Key* may read as more quotidian than fantastical to future generations, or M'Pherson recognized that the fairy tales of science would need constant updating as technological advances caught up with their fantastical prefigurings, Low emphasized how quickly the modern world was moving, redefining definitions of the possible and impossible. As with the modern world, so too with modern science, a field:

so vast, representing, on the one hand, distances running into many billions of miles, and on the other many billionths of an inch, that we are apt to become confused in trying to picture the discoveries of which we read.... A fraction of an inch relative to a billion miles is a comparison rather beyond human imagination.[402]

From such unfathomable distances to new theories of travelling faster than light, bending space and time, and cats that were neither dead or alive, the ways of managing the relationship between fantastical imaginings and accurate science worked out through these introductory scientific books were to be key in helping respond to the challenges of the new physics: quantum theory and relativity.[403] Just as scholars have identified nineteenth-century children's works such as *Alice* as key to the development of literary modernism, so too can we trace connections between the modes of comprehending and explaining new twentieth-century sciences and the literary choices, mechanisms, and tropes developed in Victorian scientific writing for young audiences.[404] In the twentieth century the contingency and strangeness of the universe were to be celebrated and communicated, science and 'research' 'no affairs of water-tight compartments and of unlimited measurings. We do not truly know any fact.'[405] As Alice's trip to Wonderland had revealed, logic could indeed now be stranger than imagination, truth stranger than fiction.

FURTHER READING

Bown, Nicola, *Fairies in Nineteenth-Century Art and Literature* (Cambridge: Cambridge University Press, 2001; 2006 paperback edition). Insightful discussion of the cultural place and image of fairies in the nineteenth century, with particular reference to works of art as well as books and periodicals.

Cosslett, Tess, *Talking Animals in British Children's Fiction, 1786–1914* (Aldershot: Ashgate, 2006). Survey of many influential children's works of the long nineteenth century that used animal characters, and includes a chapter analysing parables and fairy tales.

Fyfe, Aileen, and Lightman, Bernard (eds), *Science in the Marketplace: Nineteenth-Century Sites and Experiences* (Chicago: University of Chicago Press, 2007). Collection of articles on popular engagements with the sciences in Victorian Britain, including books, museums, lectures, and conversations.

Layton, David, *Science for the People: The Origins of the School Science Curriculum in England* (London: Allen & Unwin, 1973) Classic account of the history of science education in the nineteenth century.

Lightman, Bernard, *Victorian Popularizers of Science* (Chicago: University of Chicago Press, 2007). Extensive and knowledgeable discussion of Victorian writings for wider audiences, including in-depth analyses of publication and authorship.

O'Connor, Ralph (ed.), *Science as Romance* (London: Pickering & Chatto, 2012), Vol. VII in Gowan Dawson and Bernard Lightman (general eds), *Victorian Science and Literature* (London: Pickering & Chatto,

2011–12). Anthology of extracts from many of the texts explored here, with erudite editorial introductions.

Shuttleworth, Sally, *The Mind of the Child: Child Development in Literature, Science, and Medicine, 1840–1900* (Oxford: Oxford University Press, 2010). An exemplary analysis of childhood as represented in Victorian novels and autobiographies, medical and psychological treatises; and in the traffic between these different types of sources.

Sleigh, Charlotte, *Literature and Science* (Basingstoke: Palgrave Macmillan, 2011). An elegant introduction to literature and science scholarship.

Sumpter, Caroline, *The Victorian Press and the Fairy Tale* (Basingstoke: Macmillan, 2008; 2012 paperback edition). Expert analysis of the relationship between fairy tales and nineteenth-century print culture, particularly the periodical press.

Talairach-Vielmas, Laurence, *Fairy-Tales, Natural History and Victorian Culture* (Basingstoke: Palgrave Macmillan, 2014). Recent work detailing how metaphors and images of fairies and fairy tales were put to work as part of 'a natural historical way of knowing'.

Zipes, Jack, *Victorian Fairy-Tales: The Revolt of the Fairies and Elves* (New York: Methuen, 1987; reprinted London: Routledge, 1991). Collection of significant nineteenth-century stories, with a useful introduction from this renowned fairy tale scholar.

ENDNOTES

1. Charles Dickens, *Hard Times* (London: Penguin Popular Classics, 1994 (1854)), 1–2. See Charlotte Sleigh, *Literature and Science* (Basingstoke: Palgrave Macmillan, 2011), 93–5, for discussion of 'fact' as Dickens's 'key word' in *Hard Times*, and its relationship to concerns over mechanization and, in particular, the mechanization of education.
2. Dickens, *Hard Times*, 7.
3. Michael Slater, 'Dickens in Wonderland', in Peter L. Caracciolo (ed.), *The Arabian Nights in English Literature: Studies in the Reception of The Thousand and One Nights into British Culture*, (Basingstoke: Macmillan, 1988), 130–42. For detailed studies of Dickens and fairy tales, see: Elaine Ostry, *Social Dreaming: Dickens and the Fairy Tale* (New York and London: Routledge, 2002) (pp. 52–3 for discussion of *Hard Times*, facts and fancy), and Harry Stone, *Dickens and the Invisible World: Fairy Tales, Fantasy, and Novel-Making* (London: Macmillan, 1980) (p. 16 for discussion of facts and fancy, p. 25 for the *Arabian Nights*).
4. [Anon.], 'Old and New Toys', *Punch*, 14 (1848), 76.
5. For a brief introduction to some of the issues surrounding the advent of 'useful knowledge' campaigns in the 1820s, see James A. Secord, *Visions of Science: Books and Readers at the Dawn of the Victorian Age* (Oxford: Oxford University Press, 2014), 11–14. See also Alan Rauch, *Useful Knowledge: The Victorians, Morality, and the March of Intellect* (Durham, NC and London: Duke University Press, 2001).

6. Dickens, *Hard Times*, 1.
7. Dickens, *Hard Times*, 8. Richard Owen was the leading British comparative anatomist of the day. For the relationship between Owen and Dickens, and between periodical publication and palaeontological process, see Gowan Dawson, '"By a Comparison of Incidents and Dialogue": Richard Owen, Comparative Anatomy and Victorian Serial Fiction', *19: Interdisciplinary Studies in the Long Nineteenth-Century* 11 (2010), http://www.19.bbk.ac.uk/index.php/19/issue/view/79 [accessed 15/8/14].
8. Dickens, *Hard Times*, 9.
9. See Aileen Fyfe, 'Books and the Sciences for Young Readers', in Marina Frasca-Spada and Nicholas Jardine (eds), *Books and the Sciences in History* (Cambridge: Cambridge University Press, 2000), 276–90. Laurence Talairach-Vielmas, *Science in the Nursery: The Popularization of Science in Britain and France, 1761–1901* (Newcastle: Cambridge Scholars Publishing, 2011), also collects a series of essays on science for children in this period.
10. For an overview of 'literary manifestations' of scientific writing for wider audiences, and the cultural place of the sciences in the nineteenth century, see: Gowan Dawson, 'Science and its Popularization', in Joanne Shattock (ed.), *The Cambridge Companion to English Literature, 1830–1914* (Cambridge: Cambridge University Press, 2010), 165–83.
11. Jack Morrell and Arnold Thackray, *Gentlemen of Science: Early Years of the British Association for the Advancement of Science* (Oxford: Clarendon Press, 1981); Secord, *Visions of Science*.
12. Steven Shapin and Barry Barnes, 'Science, Nature and Control: Interpreting Mechanics' Institutes', *Social Studies of Science* 7 (1977), 31–74.
13. Jonathan R. Topham, 'Science and Popular Education in the 1830s: The Role of the *Bridgewater Treatises*', *British Journal for the History of Science* 25 (1992), 397–430; 'Beyond the Common Context: The Production and Reading of the Bridgewater Treatises', *Isis* 89 (1998), 233–62.
14. [Charles Dickens], 'The Poetry of Science', *The Examiner*, 2132 (1848), 787–8, on p. 787. For juvenile audiences, see Jonathan

R. Topham, 'Periodicals and the Making of Reading Audiences for Science in Early Nineteenth-Century Britain: The Youth's Magazine, 1828–37', in Louise Henson et al. (eds) *Culture and Science in the Nineteenth-Century Media* (Aldershot: Ashgate, 2004), 57–70.

15. Secord, *Visions of Science*; Jonathan R. Topham, 'Scientific Publishing and the Reading of Science in Nineteenth-Century Britain: A Historiographical Survey and Guide to Sources', *Studies in History and Philosophy of Science* 31 (2000), 559–612.

16. James A. Secord, *Victorian Sensation: The Extraordinary Publication, Reception, and Secret Authorship of* Vestiges of the Natural History of Creation (Chicago: University of Chicago Press, 2000).

17. Ruth Barton, '"Men of Science": Language, Identity and Professionalisation in the Mid-Victorian Scientific Community', *History of Science* 41 (2003), 73–119.

18. This argument is made in particular in Nicola Bown, *Fairies in Nineteenth-Century Art and Literature* (Cambridge: Cambridge University Press, 2006).

19. Quoted in M. O. Grenby, 'Tame Fairies Make Good Teachers: The Popularity of Early British Fairy Tales', *The Lion and the Unicorn* 30 (2006), 1–24, 3.

20. A letter written by essayist Charles Lamb in 1802, as quoted in Tess Cosslett, *Talking Animals in British Children's Fiction, 1786–1914* (Aldershot: Ashgate, 2006), 27.

21. Priscilla Wakefield, *An Introduction to the Natural History and Classification of Insects* (London: Darton, Harvey & Darton, 1816), p. iii.

22. For just one example of this, see Maria Edgeworth and Richard Lovell Edgeworth, *Practical Education* (London: J. Johnson, 1798).

23. John Aiken and Anna Laetitia Barbauld, *Evenings at Home; Or, the Juvenile Budget Opened*, ed. Aileen Fyfe (Bristol: Thoemmes Press, 2003 reprint (1809 edn)). See Fyfe, 'Young Readers', for an introduction to these scientific books, and to *Evenings at Home* in particular. James A. Secord, 'Newton in the Nursery: Tom Telescope and the Philosophy of Tops and Balls, 1761–1838', *History of Science* 23 (1985), 127–51, traces the publishing history of one important early book. For discussion of one format in particular, see Greg Myers, 'Science for Women and Children: The Dialogue

of Popular Science in the Nineteenth Century', in John Christie and Sally Shuttleworth (eds), *Nature Transfigured: Science and Literature, 1700–1900* (Manchester: Manchester University Press, 1989), 171–200.

24. J. H. Plumb, 'The New World of Children in Eighteenth-Century England', *Past and Present* 67 (1975), 64–95, 88; Jill Shefrin, '"Make it a Pleasure and Not a Task": Educational Games for Children in Georgian England', *Princeton University Library Chronicle* LX, no. 2 (1999), 251–75, G. L'E. Turner, *Nineteenth-Century Scientific Instruments* (Berkeley: University of California Press, 1983).

25. [Jacob and Wilhelm Grimm], *German Popular Stories, translated from the Kinder und Haus Märchen collected by M. M. Grimm, From Oral Tradition* (London: C. Baldwyn, 1823), p. iv.

26. The cultivation of what has been characterized as 'polite' engagement with scientific topics is analysed in Alice N. Walters, 'Conversation Pieces: Science and Politeness in Eighteenth-Century England', *History of Science* 35 (1997): 121–54, and in Katie Taylor, 'Mogg's Celestial Sphere (1813): The Construction of Polite Astronomy', *Studies in History and Philosophy of Science* 40 (2009), 360–71.

27. See Sally Shuttleworth, *The Mind of the Child: Child Development in Literature, Science and Medicine, 1840–1900* (Oxford: Oxford University Press, 2010), 60–74 for further discussion of 'lies and imagination'.

28. See Grenby, 'Tame Fairies', 10, for an introduction to the 'moral fairy tale'.

29. As reproduced in Grenby, 'Tame Fairies', 15.

30. An excellent overview of this fairy tale fashionability is provided in Caroline Sumpter, 'Fairy Tale and Folklore in the Nineteenth Century', *Literature Compass* 6 (2009), 785–98. For more in-depth analyses of fairies and fairy tales in the Victorian period, see especially: Bown, *Fairies*; Diane Purkiss, *Troublesome Things: A History of Fairies and Fairy Stories* (London: Allen Lane, 2000), 220–64; Carole Silver, *Strange and Secret Peoples: Fairies and Victorian Consciousness* (Oxford: Oxford University Press, 1999); Caroline Sumpter, *The Victorian Press and the Fairy Tale* (Basingstoke:

Macmillan, 2008, 2012 paperback edition); and the works of Jack Zipes, in particular his *Victorian Fairy Tales* (New York: Methuen, 1987; reprinted London: Routledge, 1991). A useful edited collection of fairy tales is Maria Tatar (ed.) *The Classic Fairy Tales* (London and New York: Norton, 1999).

31. Disraeli privately referred to the monarch as a 'faery': see Talairach-Vielmas, *Fairy Tales*, 80.

32. Discussed in Bown, *Fairies*.

33. Silver, *Strange and Secret Peoples*, 3.

34. Zipes, *Victorian Fairy Tales*, p. xvii.

35. Zipes, *Victorian Fairy Tales*, p. xviii.

36. Zipes, *Victorian Fairy Tales*, p. xviii.

37. Zipes, *Victorian Fairy Tales*, p. xix; see also Jack Zipes, *Fairy Tales and the Art of Subversion: The Classical Genre for Children and the Process of Civilization* (London: Routledge, 2006).

38. For one recent analysis of Lang's books, drawing on his range of interests and his particular interest in facts, see Leigh Wilson, '"There the facts are": Andrew Lang, Facts and Fantasy', *Journal of Literature and Science* 6 (2013), http://www.literatureandscience.org/wp-content/uploads/2013/11/JLS-6.2-Wilson.pdf [accessed 18/8/14].

39. [Charles Dickens], 'Frauds on the Fairies', *Household Words* VIII (1853), 97–100.

40. [Edmund Ollier], 'Yule-Tide Stories. A Collection of Scandinavian and North German Popular Tales and Traditions, from the Swedish, Danish, and German', *The Athenaeum* 1322 (1853), 247–8, 247. For further discussion of this article, see Sumpter, *Victorian Press*, 178.

41. [Ollier], 'Yule-Tide Stories', 247.

42. See, for instance, the poem 'The Fairy-Tales of Science', in *Punch* 85 (1883), 134.

43. Margaret Gatty, *Parables from Nature* (London: Bell and Daldy, 1855), p. x.

44. [John Lindley], 'The Crystal Palace Gardens', *The Athenaeum* (1854), 780.

45. John Cargill Brough, *The Fairy-Tales of Science* (London: Griffin and Farran, 1859), 3.

46. Revd H[enry] N. Hutchinson, *Extinct Monsters: A Popular Account of Some of the Larger Forms of Ancient Animal Life* (London: Chapman & Hall, 1897, 5th edn (1892)).

47. Brough, *Fairy-Tales*. See, for instance, Deborah Cadbury, *The Dinosaur Hunters: A Story of Scientific Rivalry and the Discovery of the Prehistoric World* (London: Fourth Estate, 2000), for a readable introduction to the contemporary interest in palaeontology.

48. Misquoted as the epigraph to Brough, *Fairy-Tales*, [p. i].

49. Just some of these works include: Revd T. Wilson, *The Little Geologist: or First Book of Geology* (London: Darton & Clarke [1830]); Gideon Mantell, *Thoughts on a Pebble: or, A First Lesson in Geology* (London: Reeve, Benham & Reeve, 1836 and subsequent editions); John Mill, *The Fossil Spirit: a Boy's Dream of Geology* (London: Darton & Co., 1854).

50. Brough, *Fairy-Tales*, [p. iii].

51. See discussion of the interplay between the illustration and the text, in O'Connor, *Science as Romance*, 45-6.

52. Brough, *Fairy-Tales*, 1-4.

53. Brough, *Fairy-Tales*, 4.

54. For discussion of Lyell's *Principles*, see James A. Secord's introduction to Charles Lyell, *Principles of Geology* (London: Penguin, 1997 (1830-3)).

55. Brough, *Fairy-Tales*, 4.

56. See Rudwick, *Scenes from Deep Time*, for an accessible introduction to the history of envisioning vanished worlds.

57. Brough, *Fairy-Tales*, 4, 6.

58. For discussion of wonder as the starting-point for gaining knowledge, see Verity Hunt, 'Raising a Modern Ghost: The Magic Lantern and the Persistence of Wonder in the Victorian Education of the Senses', *Romanticism and Victorianism on the Net* 52 (2008), http://id.erudit.org/iderudit/019806ar [accessed 17/8/14].

59. Brough, *Fairy-Tales*, 4.

60. See Rudwick, *Scenes from Deep Time*, for an introduction to the practices and imagery of early nineteenth-century palaeontology.

61. See Adelene Buckland, *Novel Science: Fiction and the Invention of Nineteenth-Century Geology* (Chicago: University of Chicago Press,

2013) for a bravura discussion of nineteenth-century geology as literary practice.

62. Brough, *Fairy-Tales*, 14.

63. Brough, *Fairy-Tales*, 14.

64. For a detailed assessment of the relationship between dragons and dinosaurs in the nineteenth century, see John McGowan-Hartmann, 'Shadow of the Dragon: The Convergence of Myth and Science in Nineteenth Century Paleontological Imagery', *Journal of Social History* 47 (2013), 47–70. Ralph O'Connor also discusses the use of dragon terminology, amongst other ways of describing prehistoric creatures, in his analysis of 'Victorian Saurians: The Linguistic Prehistory of the Modern Dinosaur', *Journal of Victorian Culture* 17 (2012), 492–504.

65. Brough, *Fairy-Tales*, 5.

66. Brough, *Fairy-Tales*, 5.

67. Brough, *Fairy-Tales*, 8–9.

68. Brough, *Fairy-Tales*, 9.

69. Brough, *Fairy-Tales*, 10.

70. See Richard D. Altick, *The Shows of London* (Cambridge, Mass.: Belknap Press of Harvard University Press, 1978) for the classic analysis of the whole range of such shows; Ralph O'Connor, *The Earth on Show: Fossils and the Poetics of Popular Science, 1802–1866* (Chicago: University of Chicago Press, 2007), looks in more detail at geological panoramas.

71. O'Connor, *Earth on Show*, 263–322.

72. Gideon Mantell, *The Wonders of Geology* (London: Relfe and Fletcher, 1838), frontispiece; Mill, *Fossil Spirit*, frontispiece.

73. See, for instance, W. J. T. Mitchell, *The Last Dinosaur Book: The Life and Times of a Cultural Icon* (Chicago: University of Chicago Press, 1998), 86–92, as well as McGowan-Hartmann, 'Shadow of the Dragon'.

74. Quoted in Mitchell, *The Last Dinosaur Book*, 87–8: 'As late as 1755, Samuel Johnson's Dictionary expressed a "lingering belief in the dragon as a subspecies of reptiles" by cautiously declaring that they are "perhaps imaginary".'

75. Brough, *Fairy-Tales*, 3.

76. For further research on fairies and folklore, see especially Silver, *Strange and Secret Peoples*.

77. Lizanne Henderson, 'The Natural and Supernatural Worlds of Hugh Miller', in Lester Borley (ed.), *Celebrating the Life and Times of Hugh Miller: Scotland in the Early 19th Century: Ethnography and Folklore, Geology and Natural History, Church and Society* ([Aberdeen]: Cromarty Arts Trust & the Elphinstone Institute of the University of Aberdeen, 2003), 89–98.

78. Melanie Keene, '"An active nature": Robert Hunt and the Genres of Science-Writing', in Ben Marsden, Hazel Hutchison, and Ralph O'Connor (eds), *Uncommon Contexts: Encounters between Science and Literature, 1800–1914* (Pickering & Chatto, 2013), 39–53.

79. Hutchinson, *Extinct Monsters*, 1–2.

80. [Sala and Wills], 'Fairy-land in 'Fifty Four', *Household Words* 8 (1853), 313–17, 313.

81. Jan Piggott, *The Palace of the People: The Crystal Palace at Sydenham, 1854–1936* (London: C. Hurst, 2004). The 'geological illustrations' are discussed on pp. 158–64.

82. For more on the monster models, see: Peter Doyle and Eric Robinson, 'The Victorian "Geological Illustrations" of Crystal Palace Park', *Proceedings of the Geologists' Association* 104 (1993), 181–94; Steve McCarthy, *The Crystal Palace Dinosaurs: The Story of the World's First Prehistoric Sculptures* (London: Crystal Palace Foundation, 1994); M. J. S. Rudwick, *Scenes from Deep Time: Early Pictorial Representations of the Prehistoric World* (Chicago: University of Chicago Press, 1992); James A. Secord, 'Monsters at the Crystal Palace', in Soraya de Chadarevian and Nick Hopwood (eds) *Models: The Third Dimension of Science* (Stanford: Stanford University Press, 2004), 138–69.

83. See, especially, discussion of the three-dimensionality of the models in Secord, 'Monsters at the Crystal Palace'.

84. 'Hard Times' was serialized in *Household Words* from April to August 1854.

85. Dickens, *Hard Times*, 19; Sala and Wills, 'Fairy-land in 'Fifty Four'.

86. [Henry Morley], 'Our phantom ship on an antediluvian cruise', *Household Words* III (1851), 492–6.

87. [Sala and Wills], 'Fairyland in 'Fifty Four', 313.
88. See discussion in the introduction on 'Useful Knowledge'.
89. [Sala and Wills], 'Fairyland in 'Fifty Four', 313.
90. [Sala and Wills], 'Fairyland in 'Fifty Four', 313.
91. [Sala and Wills], 'Fairyland in 'Fifty Four', 313.
92. [Sala and Wills], 'Fairyland in 'Fifty Four', 314.
93. [Sala and Wills], 'Fairyland in 'Fifty Four', 316.
94. [Sala and Wills], 'Fairyland in 'Fifty Four', 316–17.
95. For more on this new display as part of a bigger pageant of time and space, see Nancy Rose Marshall, '"A Dim World, Where Monsters Dwell": The Spatial Time of the Sydenham Crystal Palace Dinosaur Park', *Victorian Studies* 49 (2007), 286–301.
96. See, especially, McCarthy, *Crystal Palace Dinosaurs*; Secord, 'Monsters at the Crystal Palace'.
97. [Sala and Wills], 'Fairyland in 'Fifty Four', 317.
98. [Sala and Wills], 'Fairyland in 'Fifty Four', 317.
99. [Sala and Wills], 'Fairyland in 'Fifty Four', 317.
100. [Sala and Wills], 'Fairyland in 'Fifty Four', 317.
101. [M. Oliphant], 'Modern Light Literature—Science', *Blackwood's Edinburgh Magazine* 78 (1855), 215–30, 225. See O'Connor, introduction to *Science as Romance*, for further discussion.
102. [M. Oliphant], 'Modern Light Literature', 226.
103. [M. Oliphant], 'Modern Light Literature', 226.
104. [John Leech], 'A Visit to the Antediluvian Reptiles at Sydenham—Master Tom Strongly Objects to Having His Mind Improved', *Punch* 28 (1854), [p. viii].
105. Quoted in Grenby, 'Tame Fairies', 8.
106. Grenby, 'Tame Fairies', 8. This story was published in *The Lilliputian Magazine*, a juvenile periodical.
107. [George Du Maurier], 'A Little Christmas Dream', *Punch* 55 (1868), 272.
108. See discussion in Rudwick, *Scenes from Deep Time*, 214.
109. Louis Figuier, *World before the Deluge* (New York: D. Appleton, 1866), 1.
110. [Andrew Lang and May Kendall], *That Very Mab* (London: Longmans, Green, and Co., 1885), 19–33.

111. See the discussion on insects in Bown, *Fairies*, 125–35.
112. Bown, *Fairies*, 6.
113. 'Acheta Domestica' [L. M. Budgen], *Episodes of Insect Life*, Vols I–III (London: Reeve, Benham and Reeve, 1849–51).
114. See the introduction in O'Connor, *Science as Romance*, 185–8.
115. For O'Connor, these vignettes figure 'the double vision of a simile' (O'Connor, *Science as Romance*, 187).
116. 'We Challenge all Nations!', in 'Acheta Domestica', *Episodes of Insect Life*, III. 333.
117. 'Acheta Domestica' [L. M. Budgen], *Episodes of Insect Life, revised and edited by J. G. Wood* (London: Bell and Daldy, 1867), pp. v–vi.
118. [Budgen], *Episodes of Insect Life*, I, p. vii.
119. See David Allen, 'Tastes and Crazes', in Nicholas Jardine, James A. Secord, and Emma C. Spary (eds), *Cultures of Natural History* (Cambridge: Cambridge University Press, 1996), 394–407, for an introduction to the changing fashions for particular natural historical pursuits across the nineteenth century.
120. See Sam George, 'Animated Beings: Enlightenment Entomology for Girls', *Journal for Eighteenth-Century Studies* 33 (2010), 487–505 for discussion of earlier attempts to introduce young audiences to the study of natural history: George discusses Priscilla Wakefield in particular, including her *Introduction to... Insects*, whose preface had lauded the triumph of 'reason' over 'nonsense' (see Introduction).
121. [Budgen], *Episodes of Insect Life*, I, pp. ix–x.
122. [Budgen], *Episodes of Insect Life*, III, pp. vi–vii.
123. Gatty, *Parables from Nature*, pp. vii–viii. For more on Gatty's *Parables*, see Alan Rauch, 'Parables and Parodies: Margaret Gatty's Audiences in the Parables from Nature', *Children's Literature* 25 (1997), 137–53.
124. Gatty, *Parables from Nature*, p. ix.
125. Hans Christian Andersen, 'The Butterfly' (1861), in Hans Christian Andersen, *The Complete Fairy Tales and Stories* (translated by Erik Christian Haugaard) (London: Victor Gollancz, 1974), 782–4.
126. 'Acheta Domestica', *Episodes of Insect Life, revised and edited by J. G. Wood*, 122–3.

127. L. Pasley and M. S. Pasley, *The Adventures of Madalene and Louisa: Pages from the Album of L. and M. S. Pasley, Victorian Entomologists* (London: Random House, 1982). I am grateful to Elizabeth Hale for this reference.

128. Pasleys, *Adventures*, [31].

129. Quoted in Michael A. Salmon, *The Aurelian Legacy: British Butterflies and their Collectors* (Colchester: Harley Books, 2000), 25.

130. For an earlier discussion of scrapbooking practices in relation to the natural sciences, see James A. Secord, 'Scrapbook Science: Composite Caricatures in Late Georgian England', in Ann B. Shteir and Bernard Lightman (eds), *Figuring it Out: Science, Gender, Visual Culture* (Hanover and London: University Press of New England, 2007), 164–91.

131. For some wonderful examples of these playful combinations of illustrations and photographs, often featuring elements of the natural world, including insects, see Elizabeth Siegel (ed.), *Playing with Pictures: The Art of Victorian Photocollage* (New Haven: Yale University Press, 2009).

132. Pasleys, *Adventures*, [35].

133. Madalene Pasley, *A Selection of British Butterflies and Moths* (1862), American Philosophical Society, http://amphilsoc.org/mole/view?docId=ead/Mss.595.78.P26-ead.xml [accessed 17/8/14].

134. See O'Connor, *Science as Romance*, 261.

135. 'A.L.O.E.' [Charlotte M. Tucker] *Fairy Frisket; or, Peeps at Insect Life* (London: T. Nelson and Sons, 1874), 5.

136. Discussed in Purkiss, *Troublesome Things*, 223.

137. Charles Kingsley, *Madam How and Lady Why* (London: Macmillan, 1880 (1870 Bell & Daldy, first edn)).

138. 'A.L.O.E.' [Charlotte M. Tucker], *Fairy Know-a-Bit: or, a Nutshell of Knowledge* (London: T. Nelson and Sons, 1872), 11–12.

139. Cosslett, *Talking Animals*, 37–61 for discussion of 'fabulous histories and papillonades'.

140. 'A.L.O.E.', *Fairy Frisket*, 11, 13.

141. Lucy Rider Meyer, *Real Fairy Folks; Or, Fairyland of Chemistry: Explorations in the World of Atoms* (Boston: D. Lothrop, 1887).

142. Chapter 3 was titled 'Cousins amongst the fairies': Meyer, *Real Fairy Folks*, 70–105.

143. Meyer, *Real Fairy Folks*, 68, 67.

144. Meyer, *Real Fairy Folks*, 62.

145. For this story, see Christoph Meinel, 'Molecules and Croquet Balls', in de Chadarevian and Hopwood, *Models*, 242–75.

146. Meyer, *Real Fairy Folks*, 61.

147. Meyer, *Real Fairy Folks*, 'A Word'. See Melanie Keene, 'From Candles to Cabinets: "Familiar Chemistry" in Early Victorian Britain', *Ambix* 60 (2013), 54–77, for a more extensive discussion of the use of everyday objects in elementary chemical experiments in the mid-nineteenth century.

148. Meyer, *Real Fairy Folks*, 208.

149. George Henry Lewes, *Sea-Side Studies at Ilfracombe, Tenby, The Scilly Isles and Jersey* (London and Edinburgh: William Blackwood and Sons, 1858), 35.

150. Agnes Catlow, *Drops of Water: Their Marvellous and Beautiful Inhabitants Displayed by the Microscope* (London: Reeve and Benham, 1851).

151. Catlow, *Drops of Water*, pp. x–xi.

152. Catlow, *Drops of Water*, pp. xi.

153. Catlow, *Drops of Water*, p. xi.

154. Catlow, *Drops of Water*, pp. xi–xvi.

155. Catlow, *Drops of Water*, p. xvi.

156. For a rather dated but still helpful overview, see Charles Pouthas, 'The Revolutions of 1848', in J. P. T. Bury (ed.), *The New Cambridge Modern History*, Vol. 10: *The Zenith of European Power, 1830–70* (Cambridge: Cambridge University Press, 1960), 389–412.

157. Andersen, 'The Drop of Water' (1848): http://hca.gilead.org.il/drop_wat.html [accessed 15/8/14]. A different translation is included in Andersen, *Complete Fairy Tales*, 354–5, as 'A Drop of Water'.

158. Andersen, 'The Drop of Water'.

159. Andersen, 'The Drop of Water'.

160. See Stella Butler, R. H. Nuttall, and Olivia Brown, *The Social History of the Microscope* (Cambridge: Whipple Museum, 1986)

for a helpful introduction. Kate Flint, *The Victorians and the Visual Imagination* (Cambridge: Cambridge University Press, 2000) contains much erudite discussion of the particular significance of microscopy in nineteenth-century culture.

161. Philip Henry Gosse, *Evenings at the Microscope* (London: Society for Promoting Christian Knowledge, [1859]); Wood, *Common Objects of the Microscope*.

162. See Altick, *Shows of London*, for the pioneering analysis and description of these shows. Scientific shows are particularly discussed in 'The Two Faces of Science', 363–74.

163. Altick, *Shows of London*, 369–71 for discussion of the oxy-hydrogen microscope, and reports from a demonstration of the creatures lurking within drinking water.

164. Revd J. G. Wood, *Common Objects of the Microscope* (London: Routledge, Warne and Routledge, 1861), 23.

165. For discussion of Andersen's interest in the sciences, see Ane Grum-Schwensen, 'Little Hans Christian and Great Hans Christian: The Poet and the Scientist', *Interdisciplinary Science Reviews* 30 (2005), 349–55.

166. Dennis R. Dean, *Gideon Mantell and the Discovery of Dinosaurs* (Cambridge: Cambridge University Press, 1999), 216.

167. Mantell, *Thoughts on a Pebble*; Gideon Mantell, *Thoughts on Animalcules; or, a glimpse at the invisible world revealed by the microscope* (London: John Murray, 1846).

168. Mantell, *Thoughts on Animalcules*, 1–2.

169. Wood, *Common Objects of the Microscope*, 113.

170. Catlow, *Drops of Water*, [p. vii].

171. Catlow, *Drops of Water*, [p. vii].

172. Christopher Hamlin, *A Science of Impurity: Water Analysis in Nineteenth-Century Britain* (Berkeley: University of California Press, 1990).

173. Quoted in Hamlin, *Science of Impurity*, 99.

174. This point is made in detail in Ursula Seibold-Bultmann, 'Monster Soup: The Microscope and Victorian Fantasy', *Interdisciplinary Science Reviews* 25 (2000), 211–19.

175. [Anon.], 'The Wonders of a London Water Drop', *Punch* 18 (1850), 188.

176. [Anon.], 'London Water Drop'.

177. [Henry Morley], 'The water-drops: a fairy-tale', *Household Words* I (1850), 482–9.

178. [Morley], 'The water-drops', 483.

179. [Morley], 'The water-drops', 482.

180. [Morley], 'The water-drops', 483.

181. [Morley], 'The water-drops', 484.

182. Wood, *Common Objects of the Microscope*, 112.

183. Arabella Buckley, *The Fairy-Land of Science* (London: Edward Stanford, 1879), 12–13.

184. Buckley, *Fairy-Land*, 73–98, 74.

185. Buckley, *Fairy-Land*, 5.

186. Emily Winterburn, 'The Herschels: A Scientific Family in Training', unpublished PhD thesis (Imperial College London, 2011), 191.

187. Annie Carey, *Autobiographies of a Lump of Coal; A Grain of Salt; A Drop of Water; A Bit of Old Iron; A Piece of Flint* (London: Cassell, Petter and Galpin, [1870]). For an introduction to these types of autobiographical works purportedly written by things themselves, which literary scholars have termed 'it-narratives', see Mark Blackwell, 'The It-Narrative in Eighteenth-Century England: Animals and Objects in Circulation', *Literature Compass* 1 (2004), 1–5. By the nineteenth century it-narratives were a staple of children's literature.

188. For particular discussion of the 'Voices of Nature' trope, see Lightman, '"Voices of Nature": Popularizers of Victorian Science', in Lightman (ed.) *Victorian Science in Context*, 187–211.

189. Carey, *Autobiographies*, 26.

190. Carey, *Autobiographies*, 9–10.

191. *The Times*, Wednesday, 7 December 1870, 8.

192. [Geraldine Jewsbury], 'Books for the Young', *The Athenaeum* 2289 (1871), 336–7, 336.

193. 'M.S.', 'Down the microscope and what Alice found there', *Brighter Biochemistry*, December 1927, 36–9.

194. There is a wealth of scholarship on the *Alice* books. See Martin Gardner (ed.), *The Annotated Alice* (London: Penguin, 2001) for an introduction to the deeper layers of both *Wonderland* and *Looking-Glass*.
195. 'M.S.', 'Down the microscope', 39.
196. 'M.S', 'Down the microscope', 36.
197. Charles Kingsley, *The Water-Babies: A Fairy Tale for a Land-Baby* (London: Penguin Popular Classics, 1995 (1863)).
198. 'M.S.', 'Down the microscope', 38.
199. 'M.S.', 'Down the microscope', 39.
200. 'M.S.', 'Down the microscope, 39.
201. May Kendall, *Dreams to Sell* (London: Longmans, Green, & Co., 1887), 32. Two of Kendall's poems are included in Barbara T. Gates (ed.), *In Nature's Name: An Anthology of Women's Writing and Illustration, 1780–1930* (Chicago: University of Chicago Press, 2002), 512–14. For an introduction to Kendall's biography, see Catherine Elizabeth Birch, 'Evolutionary Feminism in Late-Victorian Women's Poetry: Mathilde Blind, Constance Naden and May Kendall' (Unpublished PhD thesis, University of Birmingham, 2011), 56–65: http://etheses.bham.ac.uk/3024/1/Birch11PhD.pdf [accessed 14/8/14].
202. [Anon.], 'Treading on the Fairies' Tales', *Punch* 76 (1879), 24.
203. See Isobel Armstrong, *Victorian Glassworlds: Glass Culture and the Imagination 1830–1880* (Oxford: Oxford University Press, 2008), 205, for discussion of Perrault's (deliberate, she argues) introduction of glass as 'the founding element of the story', which can then mediate its later metamorphoses, as she argues from pp. 204–21. Talairach-Vielmas echoes Armstrong's analysis in *Fairy Tales*, 85–6.
204. [Anon.], 'Treading on the Fairies' Tales'.
205. See especially Bernard Lightman, 'Evolution for Young Victorians', *Science and Education* 21 (2012), 1015–34, for analyses of many of these works. The classic analysis of Darwin's evolutionary writing and its connections to contemporary literature is Gillian Beer, *Darwin's Plots: Evolutionary Narrative in Darwin,*

George Eliot and Nineteenth-Century Fiction (Cambridge: Cambridge University Press, 2000 (1983)).

206. This is discussed in detail in Lightman, 'Evolution for Young Victorians', 1023–32, as well as in Talairach-Vielmas, *Fairy Tales*, 15–46.

207. [Wendell Phillips Garrison], *What Mr Darwin Saw* (New York: Harper & Brothers, 1880 (1879)). See Lightman, 'Evolution for Young Victorians', 1018–23 for discussion of this text.

208. Arabella Buckley, *Life and her Children: glimpses of animal life from the amoeba to the insects* (London: E. Stanford, 1880); Buckley, *Winners in Life's Race: or, the great backboned family* (London: Edward Stanford, 1882). For just some of the scholarly work on Buckley, see: Barbara T. Gates, 'Revisioning Darwin with Sympathy: Arabella Buckley', in Gates and Ann B. Shteir (eds.), *Natural Eloquence: Women Reinscribe Science* (Madison: University of Wisconsin Press, 1997), 164–76; Bernard Lightman, *Victorian Popularizers of Science: Designing Nature for New Audiences* (Chicago: University of Chicago Press, 2007), 239–53 for discussion of Buckley and the 'evolutionary epic'; Talairach-Vielmas, *Fairy Tales*, 47–64; Richard Somerset, 'Bringing (Anti-)Evolutionism into the Nursery: Narrative Strategies in the Emergent "History of Life" Genre', in Talairach-Vielmas (ed.), *Science in the Nursery*, 140–63.

209. An accessible introduction to the evolutionary context and content of Kingsley's work can be found in John Beatty and Piers J. Hale, '*Water Babies*: An Evolutionary Parable', *Endeavour* 32 (2008), 141–6; or in more depth in Piers J. Hale, 'Monkeys into Men and Men into Monkeys: Chance and Contingency in the Evolution of Man, Mind and Morals in Charles Kingsley's *Water Babies*', *Journal of the History of Biology* 46 (2013), 551–97. Jessica Staley's insightful analysis in 'Of Beasts and Boys: Kingsley, Spencer, and the Theory of Recapitulation', *Victorian Studies* 49 (2007), 583–609, elucidates how the book engages with Spencerian evolutionary theories, testing natural, literary, and moral systems of education. She concludes that Kingsley argues for the pedagogic efficacy of fantastic literature, rather than the

merely realistic. Christopher Hamlin's recent article argues that an 'ecocritical' approach can unite Kingsley's disparate interests ('Charles Kingsley: From Being Green to Green Being', *Victorian Studies* 54 (2012), 255–81).

210. Kingsley, *Water-Babies*.

211. Kingsley, *Water-Babies*, 70. For discussion of these didactic dialogues, see Myers, 'Science for Women and Children', and Michèle Cohen, '"Familiar Conversation": The Role of the "Familiar Format" in Education in Eighteenth- and Nineteenth-Century England', in Mary Hilton and Jill Shefrin (eds), *Educating the Child in Enlightenment Britain: Beliefs, Cultures, Practices* (Aldershot: Ashgate, 2009), 99–116.

212. Kingsley, *Water-Babies*, 59–60.

213. Kingsley, *Water-Babies*, 74.

214. See Cosslett, *Talking Animals*, 121–2.

215. For discussion of Buckley, in particular, see Richard Somerset, 'Arabella Buckley and the Feminization of Evolution as a Communication Strategy', *Nineteenth-Century Gender Studies* 7 (2011): http://www.ncgsjournal.com/issue72/issue72.htm [accessed 15/8/14].

216. Albert and George Gresswell, *The Wonderland of Evolution* (London: Field & Tuer, [1884]). See Lightman, 'Evolution for Young Victorians', 1028–30, and O'Connor, *Science as Romance*, 215–18, for discussion of this book.

217. Gresswells, *Wonderland*, 4 for 'lowly form'.

218. Charles Kingsley, *Alton Locke, Tailor and Poet: An Autobiography* (Oxford: Oxford University Press, 1983 (1850)), 334–50.

219. Gresswells, *Wonderland*, 131–2. See Lightman, 'Evolution for Young Victorians', for the Gresswells' veterinary expertise.

220. For a good introduction to and overview of these evolutionary debates, see Peter Bowler, *Evolution: The History of an Idea* (Berkeley: University of California Press, 2003), especially chapters 6–8.

221. Gresswells, *Wonderland*, 4, 5.

222. Gresswells, *Wonderland*, 5.

223. Gresswells, *Wonderland*, 5.

224. Gresswells, *Wonderland*, 5.

225. Gresswells, *Wonderland*, 90.

226. See Cosslett, *Talking Animals*, for extensive analyses of such chatty creatures.

227. Gresswells, *Wonderland*, 93.

228. Gresswells, *Wonderland*, 65. See Charlotte Sleigh, *Ant* (London: Reaktion Books, 2003), for an introduction to the cultural history of ants and their use throughout history to comment on the structure and relations of society.

229. Gresswells, *Wonderland*, 69.

230. Gresswells, *Wonderland*, 69–70.

231. Armstrong comments on the particular use of elastic glass in Victorian tellings of the Cinderella story—a magical object that transcended its sandy origins to become associated with female bodies and reproduction (Armstrong, *Victorian Glassworlds*, 206–8).

232. Gresswells, *Wonderland*, 79.

233. [Anon.], '"The Wonderland of Evolution"', *Glasgow Herald*, Thursday, 9 October 1884, 3.

234. [Anon.], '"The Wonderland of Evolution"', *Glasgow Herald*, Thursday, 9 October 1884, 3.

235. Gresswells, *Wonderland*, 46.

236. [Anon.], '"The Wonderland of Evolution"', *Glasgow Herald*, Thursday, 9 October 1884, 3.

237. [Anon.], '"The Wonderland of Evolution"', *Glasgow Herald*, Thursday, 9 October 1884, 3.

238. [Anon.], '"The Wonderland of Evolution"', *The Morning Post*, Tuesday, 30 September 1884, 2.

239. [Anon.], '"The Wonderland of Evolution"', *The Morning Post*, Tuesday, 30 September 1884, 2.

240. [Anon.], '"The Wonderland of Evolution"', *The Morning Post*, Tuesday, 30 September 1884, 2.

241. Lightman, 'Evolution for Young Victorians', 1030.

242. Gresswells, *Wonderland*, 131.

243. Quoted in Sumpter, *Victorian Press*, 41. Also discussed in Talairach-Vielmas, *Fairy Tales*, 153.

244. Kingsley, *Madam How*, 130.

245. Kingsley, *Madam How*, 140, 142.
246. Lilian Gask, In the '*Once Upon a Time*': a fairy-tale of science (London: George G. Harrap & Co., [1913]).
247. Gask, '*Once Upon a Time*', 5.
248. Gask, '*Once Upon a Time*', 5–6.
249. Lilian Gask, In *Nature's School* (London: George G. Harrap, [1908]).
250. See Sally Gregory Kohlstedt, *Teaching Children Science: Hands-On Nature Study in North America, 1890-1930* (Chicago: University of Chicago Press, 2010), for an in-depth analysis of the 'Nature Study' movement.
251. Gask, '*Once Upon a Time*', 14.
252. Gask, '*Once Upon a Time*', 22–3.
253. Gask, '*Once Upon a Time*', 35.
254. For discussion of the new science museums of the period, see Sophie Forgan, 'The Architecture of Display: Museums, Universities and Objects in Nineteenth-Century Britain', *History of Science* 32 (1994), 139–62, and Carla Yanni, *Nature's Museums: Victorian Science and the Architecture of Display* (London: Athlone, 1999).
255. Revd H. N. Hutchinson, *Prehistoric Man and Beast* (London: Smith, Elder & Co., 1896).
256. Gask, '*Once Upon a Time*', 29–30.
257. Gask, '*Once Upon a Time*', 118.
258. Gask, '*Once Upon a Time*', 111.
259. Gask, '*Once Upon a Time*', 282–3.
260. Kendall, *Dreams to Sell*, 14–16.
261. Kendall, *Dreams to Sell*, 23–5.
262. Kendall, *Dreams to Sell*, 25.
263. Kendall, *Dreams to Sell*, 31–2. See Chapter 2 for discussion of Kendall and Lang's novel, *That Very Mab*.
264. Kendall, *Dreams to Sell*, 31.
265. Kendall, *Dreams to Sell*, 31, 32.
266. Kendall, *Dreams to Sell*, 31.
267. [Harriet Martineau], *Harriet Martineau's Autobiography with Memorials by Maria Weston Chapman*, (London: Smith, Elder & Co., 1877 third edition), 15.
268. [Martineau], *Autobiography*, 20.

269. The most famous of these ghostly apparitions was Pepper's Ghost, discussed in Chapter 6.

270. Arabella B[urton] Buckley, *Through Magic Glasses, and other lectures: a sequel to 'The Fairy-land of Science'* (London: Edward Stanford, 1890).

271. Buckley, *Through Magic Glasses*, preface, [1].

272. Buckley, *Through Magic Glasses*, 5.

273. David Brewster, *Letters on Natural Magic, addressed to Sir Walter Scott* (London: John Murray, 1832). This work was explicitly written as a riposte to Scott's *Letters on Demonology*. Brewster's work appeared in John Murray's new 'family library' as part of a new series of works published for this domestic market. See Scott Bennett, 'John Murray's Family Library and the Cheapening of Books in Early Nineteenth-Century Britain', *Studies in Bibliography* 29 (1976), 139–66. For further discussion of Brewster, see Hunt, 'Raising a Modern Ghost'.

274. Buckley, *Through Magic Glasses*, 2.

275. Buckley, *Through Magic Glasses*, 3.

276. Buckley, *Through Magic Glasses*, 4.

277. Buckley, *Through Magic Glasses*, 28.

278. Buckley, *Through Magic Glasses*, 29.

279. Buckley, *Through Magic Glasses*, 42.

280. Buckley, *Through Magic Glasses*, 52–3.

281. Buckley, *Through Magic Glasses*, 53–4.

282. Buckley, *Through Magic Glasses*, 55–6.

283. Buckley, *Through Magic Glasses*, 56.

284. Buckley, *Through Magic Glasses*, 57.

285. Buckley, *Through Magic Glasses*, 58.

286. [Michael A. Denham], *A Few Fragments of Fairyology, shewing its connection with natural history* (Civ. Dunelm.: Will. Duncan & Son, 1859), 4. This unusual work was printed in an edition of only 50 copies.

287. See Kohlstedt, *Teaching Children Science*, for descriptions of such 'nature study' lessons in actual schools.

288. Buckley, *Through Magic Glasses*, 60.

289. Buckley, *Through Magic Glasses*, 68.

290. Buckley, *Through Magic Glasses*, 73.

291. Agnes Giberne, *Among the Stars or Wonderful Things in the Sky* (New York: American Tract Society [1884]).

292. Giberne, *Among the Stars*, [p. vii]. Her previous work was Agnes Giberne, *Sun, Moon, and Stars: A Book for Beginners* (New York: Carter, 1879).

293. Lizzie W. Champney, *In the Sky-Garden* (Boston: Lockwood, Brooks, and Company, 1877), [3].

294. See: http://vcencyclopedia.vassar.edu/alumni/elizabeth-williams-champney.html [accessed 1/8/14].

295. Champney, *Sky-Garden*, 15.

296. Champney, *Sky-Garden*, 114–15.

297. Champney, *Sky-Garden*, 116.

298. Robert Louis Stevenson, *A Child's Garden of Verses* (New York: Dodge Publishing Co., 1905 (1885)), 58.

299. Quoted in Graeme Gooday, *Domesticating Electricity: Technology, Uncertainty and Gender, 1880–1914* (London: Pickering & Chatto, 2008), 107.

300. [E. L. L. Blanchard], 'Earth, Air, Fire and Water or Harlequin Gas and the Flight of the Fairies' ('Land of Light! or Harlequin Gas! and the Four Elements Earth, Air, Fire and Water!') (Victoria Theatre, dated 4 December 1848), British Library Add MS 43015, fos. 518–32. I am very grateful to Rupert Cole for providing me with this reference: see his discussion of the pantomime in 'Science Steals the Show', published in *Viewpoint: Magazine of the British Society for the History of Science* Christmas Special 2012, 2–3, 2: http://www.bshs.org.uk/wp-content/uploads/View point-Christmas-Special-2012-small.pdf [accessed 12/08/14]

301. Jim Davis, 'Introduction', in Jim Davis (ed.) *Victorian Pantomime: A Collection of Critical Essays* (Basingstoke: Palgrave Macmillan, 2010), 1–18, 5.

302. Purkiss, *Troublesome Things*, 225–31; Jeffrey Richards, 'E. L. Blanchard and "The Golden Age of Pantomime"', in Davis, *Victorian Pantomime*, 21–40, 23.

303. Janice Norwood, 'Harlequin Encore: Sixty Years of the Britannia Pantomime', in Davis, *Victorian Pantomime*, 70–84, 71.

304. Norwood, 'Harlequin Encore', 77. For discussion of how common railway accidents had become by the late 1840s, see Ralph Harrington, 'The Railway Accident: Trains, Trauma and Technological Crisis in Nineteenth-Century Britain' (University of York: Working Papers in Railway & Transport Studies, 1999): http://www.york.ac.uk/inst/irs/irshome/papers/rlyacc.htm [accessed 14/8/14].

305. [Anon.], 'Music and the Drama', *Athenaeum* 1105 (1848), 1337–9, 1338.

306. [Blanchard], 'Land of Light', 522.

307. Richards, 'Blanchard and the "Golden Age"', 26; see Clement Scott and Cecil Howard, *The Life and Reminiscences of E. L. Blanchard, with notes from the diary of Wm Blanchard*, Vol. I (London: Hutchinson & Co., 1891), 65, for confirmation of the pantomime's authorship.

308. Richards, 'Blanchard and the "Golden Age"', 26.

309. Norwood, 'Harlequin Encore', 79.

310. [Blanchard], 'Land of Light', 519.

311. [Blanchard], 'Land of Light', 519.

312. Norwood, 'Harlequin Encore', 71.

313. [Blanchard], 'Land of Light', 519. Underlining of pun in original manuscript.

314. [Blanchard], 'Land of Light', 520.

315. This reference is to the Thames Tunnel, opened in 1843, and on which two Brunels, father and son, worked. See, for instance: Alan Muir Wood, 'Brunel, Sir (Marc) Isambard (1769–1849)', *Oxford Dictionary of National Biography*, Oxford University Press, 2004, online edn May 2005 [http://www.oxforddnb.com/view/article/3774, accessed 13/8/14]; and R. Angus Buchanan, 'Brunel, Isambard Kingdom (1806–1859)', *Oxford Dictionary of National Biography*, Oxford University Press, 2004, online edn Jan. 2011 [http://www.oxforddnb.com/view/article/3773, accessed 13/8/14].

316. [Blanchard], 'Land of Light', 520.

317. [Blanchard], 'Land of Light', 521.

318. [Blanchard], 'Land of Light', 521–2.

319. Quoted in Davis, 'Introduction', 11.

320. [Blanchard], 'Land of Light', 523–32, 28.

321. [Blanchard], 'Land of Light', 528.

322. Norwood, 'Harlequin Encore', 79, 80.

323. See Iwan Morus, 'More the Aspect of Magic than Anything Natural: The Philosophy of Demonstration in Victorian Popular Science', in Lightman and Fyfe (eds), *Science in the Marketplace*, 336–70. For a geographically minded tour of other spaces of nineteenth-century science, see Bernard Lightman, 'Refashioning the Spaces of London Science: Elite Epistemes in the Nineteenth Century', in David N. Livingstone and Charles W. J. Withers (eds), *Geographies of Nineteenth-Century Science* (Chicago: University of Chicago Press, 2011), 25–50.

324. *The Era*, Vol. XI, no. 535 (24 December 1848). The Royal Polytechnic Institution's diving bell is shown in Morus, 'Aspect of Magic', 342.

325. Morus, 'Aspect of Magic', 363–6. As a young man Blanchard had toured the country giving lecture-demonstrations on the oxyhydrogen microscope and on entomology (Scott and Howard, *Life and Reminiscences*, 13–16).

326. James A. Secord, 'Quick and Magical Shaper of Science', *Science* 297 (2002), 1648–9.

327. As referenced in Secord, 'Quick and Magical'.

328. Morus, 'Aspect of Magic', 359–63.

329. Quoted in Morus, 'Aspect of Magic', 360.

330. Morus, 'Aspect of Magic', 362.

331. Brough, *Fairy-Tales*, 25. See the discussion in Gooday, *Domesticating Electricity*, 197–217 for an extended analysis of electricity's varied figuring as fairy, infant, sprite, and genie in the second half of the nineteenth century.

332. For the history of magnetism see, for instance, Patricia Fara, *Sympathetic Attractions: Magnetic Practices, Beliefs, and Symbolism in Eighteenth-Century England* (Princeton: Princeton University Press, 1996).

333. Brough, *Fairy-Tales*, 19. For this imperial context see in particular Bruce J. Hunt, 'Doing Science in a Global Empire: Cable Telegraphy and Electrical Physics in Victorian Britain', in Lightman (ed.), *Victorian Science in Context*, 312–33.

334. Brough, *Fairy-Tales*, 25.
335. Brough, *Fairy-Tales*, 25.
336. *The Children's Fairy Geography or a Merry Trip*, by Forbes E. Winslow M.A. (W. Skeffington & Son [1879?]).
337. Winslow, *Fairy Geography*, p. x.
338. This echoed the 1848 *Punch* cartoon on 'Old and New Toys', which was discussed in the Introduction.
339. Winslow, *Fairy Geography*, p. x.
340. Teresa Ploszajska, '"Cloud Cuckoo Land?" Fact and Fantasy in Geographical Readers, 1870–1944', *Paradigm* 21 (1996), unpaginated online article, given quotation taken from second paragraph. Available at: http://faculty.education.illinois.edu/westbury/paradigm/ploszajska.html [accessed 27/7/14].
341. [Anon.], 'The Children's Fairy Geography. By Forbes E. Winslow', *The Spectator*, 3 January 1880, 31–2, 31.
342. [Anon.], 'The Children's Fairy Geography', 31.
343. [Anon.], 'The Children's Fairy Geography', 31.
344. See Aileen Fyfe, 'Natural History and the Victorian Tourist: From Landscapes to Rock Pools', in Livingstone and Withers (eds), *Geographies of Nineteenth-Century Science*, 371–98.
345. Winslow, *Fairy Geography*, 16–17.
346. Winslow, *Fairy Geography*, p. x.
347. Winslow, *Fairy Geography*, 54. For more on the magical associations of the train, see Michael J. Freeman, *Railways and the Victorian Imagination* (New Haven: Yale University Press, 1999), 38.
348. Winslow, *Fairy Geography*, 21.
349. Winslow, *Fairy Geography*, 43.
350. See Richard Holmes, *Falling Upwards: How We Took to the Air* (London: William Collins, 2013) for an engaging introduction to the history of ballooning and, particularly, balloonists.
351. This point is made in Holmes, *Falling Upwards*, 52.
352. Winslow, *Fairy Geography*, 21, 22.
353. For a collection of essays on aerial perspectives, see: Mark Dorrian and Frédéric Pousin (eds), *Seeing from Above: The Aerial View in Visual Culture* (London: I.B. Tauris, 2013).
354. Bown, *Fairies*, 54, 56.

355. See Holmes, *Falling Upwards*, 61–72 for a full account of this extraordinary flight.

356. Holmes, *Falling Upwards*, 60–1.

357. Holmes, *Falling Upwards*, 65.

358. Winslow, *Fairy Geography*, 310–11.

359. Winslow, *Fairy Geography*, 131.

360. Winslow, *Fairy Geography*, 309.

361. Holmes, *Falling Upwards*, 63.

362. Winslow, *Fairy Geography*, 224.

363. Winslow, *Fairy Geography*, 130.

364. Winslow, *Fairy Geography*, 176–7.

365. L. Frank Baum, *The Master Key: An Electrical Fairy Tale* (London: B. F. Stevens & Brown, [1902] (1901)). The description is given in the book's subtitle.

366. Baum, *Master Key*, unpaginated preface.

367. Vivian Wagner, 'Unsettling Oz: Technological Anxieties in the Novels of L. Frank Baum', *The Lion and the Unicorn* 30 (2006) 25–53, 31.

368. Baum, *Master Key*, 1.

369. Baum, *Master Key*, 1.

370. Baum, *Master Key*, 3.

371. Baum, *Master Key*, 10.

372. Baum, *Master Key*, 17. The use of the word 'demon' also connotes James Clerk Maxwell's 'Demon'; his 1867 thought-experiment about the second law of thermodynamics.

373. Baum, *Master Key*, 12–13. For discussion of the origins of notions of scientific 'genius', see Simon Schaffer, 'Genius in Romantic Natural Philosophy', in Andrew Cunningham and Nicholas Jardine (eds), *Romanticism and the Sciences* (Cambridge: Cambridge University Press, 1990), 82–98.

374. Baum, *Master Key*, 128–32.

375. Baum, *Master Key*, 215–22.

376. Baum, *Master Key*, 244–5.

377. L. Frank Baum, *The Wonderful Wizard of Oz* (London: J. M. Dent & Sons, 1965 (1900)).

378. Baum, *Wizard of Oz*, 102–9.

379. Baum, *Wizard of Oz*, 102.

380. For Seth Lehrer, there is an 'inherent theatricality' to Oz, compounded by the appearance of the Wizard as 'a circus entrepreneur', modelled in later works in the series on P. T. Barnum himself (Lehrer, *Children's Literature*, 248).

381. Quoted in Bown, *Fairies*, 52. Baum, *Wizard of Oz*, 105–6, 114–18.

382. Of course, both Dorothy and the Wizard are in fact quite extraordinary, and are able to transcend their quotidian origins.

383. [James Hinton], 'The Fairy Land of Science', *Cornhill Magazine* 5/ 25 (January 1862), 36–42, 36. See discussion of Hinton's particular Kantian metaphysical commitments in Bown, *Fairies*, 99–103.

384. [Hinton], 'Fairy Land', 36.

385. Bown, *Fairies*, 98.

386. [Hinton], 'Fairy Land', 38.

387. [Hinton], 'Fairy Land', 39.

388. Low, *Science in Wonderland*.

389. Byron, *Don Juan*.

390. Brough, *Fairy-Tales of Science*, 3.

391. Buckley, *Fairy-Land of Science*, 6.

392. Buckley, *Fairy-Land of Science*, 2.

393. Kingsley, *Water-Babies*, 144–5. See discussion in Sleigh, *Literature and Science*, 148–50.

394. [Hinton], 'Fairy Land', 39.

395. See Clare Pettitt, *Patent Inventions: Intellectual Property and the Victorian Novel* (Oxford: Oxford University Press, 2004).

396. G[eorge] Gamow's *Mr Tompkins in Wonderland, or Stories of c. G, and h, Illustrated by John Hookham* (Cambridge: Cambridge University Press, 1939), was also inspired by *Alice*, and dedicated to Lewis Carroll and Niels Bohr. See Sophie Forgan, 'Atoms in Wonderland', *History and Technology* 19 (2003), 177–96.

397. Professor A[rchibald] M[ontgomery] Low, *Science in Wonderland* (London: Lovat, Dickson & Thompson Ltd, [1935]), 9.

398. Low, *Science in Wonderland*, 10.

399. Low, *Science in Wonderland*, 12.

400. Popular science writing was also by now an established genre. See Peter Bowler, *Science for All: The Popularization of Science in*

Early Twentieth-Century Britain (Chicago: University of Chicago Press, 2009).

401. Low, *Science in Wonderland*, 14–15.

402. Low, *Science in Wonderland*, 10–11.

403. For more on the relationship between the new physics of the early twentieth century and scientific, poetic, journalistic, and fictional writing, see: Katy Price, *Loving Faster Than Light: Romance and Readers in Einstein's Universe* (Chicago: University of Chicago Press, 2012), Michael H. Whitworth, *In Einstein's Wake: Relativity, Metaphor, and Modernist Literature* (Oxford: Oxford University Press, 2001).

404. See Juliet Dusinberre, *Alice to the Lighthouse: Children's Books and Radical Experiments in Art* (Basingstoke: Macmillan, 1987).

405. Low, *Science in Wonderland*, 12.

INDEX